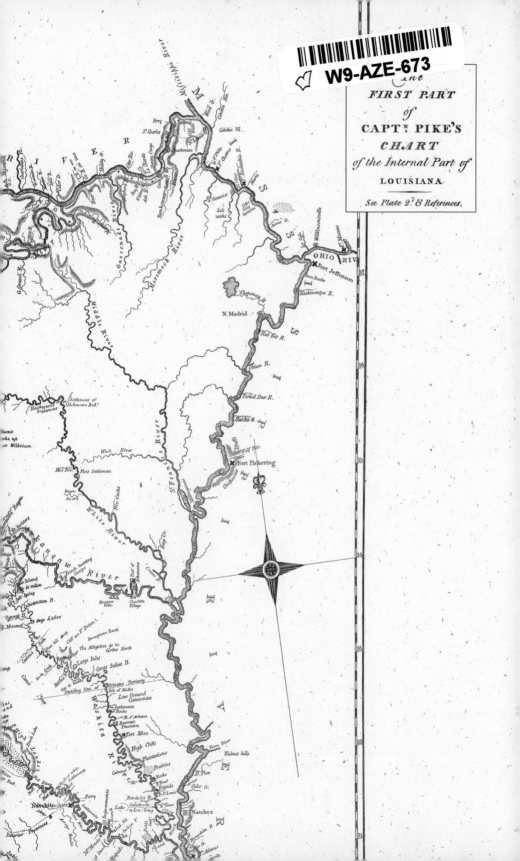

The Land between the Rivers

The Land between the Rivers

Thomas Nuttall's Ascent

of

the Arkansas, 1819

RUSSELL M. LAWSON

THE UNIVERSITY OF MICHIGAN PRESS
ANN ARBOR

For my parents

Copyright © by Russell M. Lawson 2004
All rights reserved
Published in the United States of America by
The University of Michigan Press
Manufactured in the United States of America
⊗ Printed on acid-free paper

2007   2006   2005   2004   4   3   2   1

A CIP catalog record for this book is available from the British Library.

Library of Congress Cataloging-in-Publication Data

Lawson, Russell M., 1957–
   The land between the rivers : Thomas Nuttall's ascent of the Arkansas,
1819 / Russell M. Lawson.
      p.   cm.
   Includes index.
   ISBN 0-472-11411-5 (acid-free paper)
      1. Arkansas River—Description and travel.   2. Southwest, Old—
Description and travel.   3. Nuttall, Thomas, 1786–1859—Travel—Arkansas
River.   4. Nuttall, Thomas, 1786–1859—Travel—Southwest, Old.   5. Indians
of North America—Southwest, Old—History—19th century.   6. Natural
history—Arkansas River Valley.   7. Natural history—Southwest, Old.
I. Title.
F417.A7L39   2004
917.67'3043—dc22   2003027609

Frontispiece: Portrait of Thomas Nuttall from the archives of the Gray Herbarium,
Harvard University, Cambridge, Massachusetts, USA.

Map of the Three Forks drawn by Benjamin A. Lawson after Tom Meagher,
Sketch Map of the Three Forks (Tulsa, 1942).

Endsheets from *The Expeditions of Zebulon Montgomery Pike,* edited by Elliot Coues
(New York: Harper, 1895).

Contents

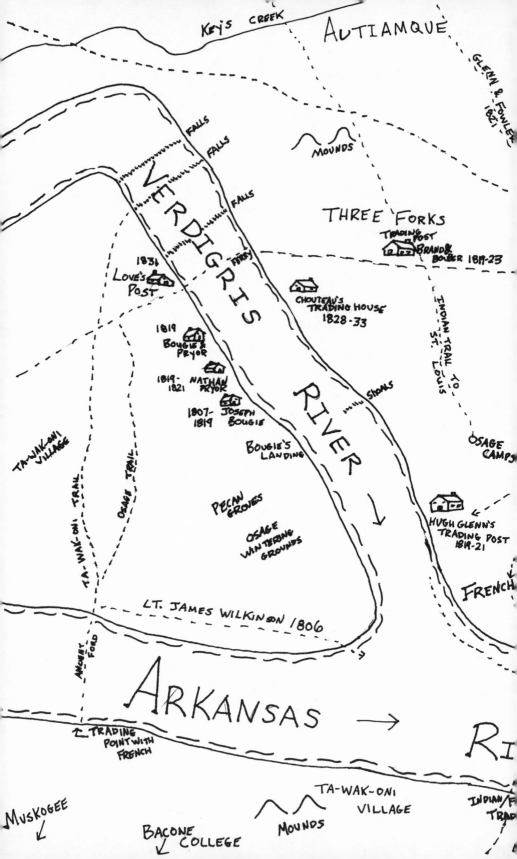

he Three Forks

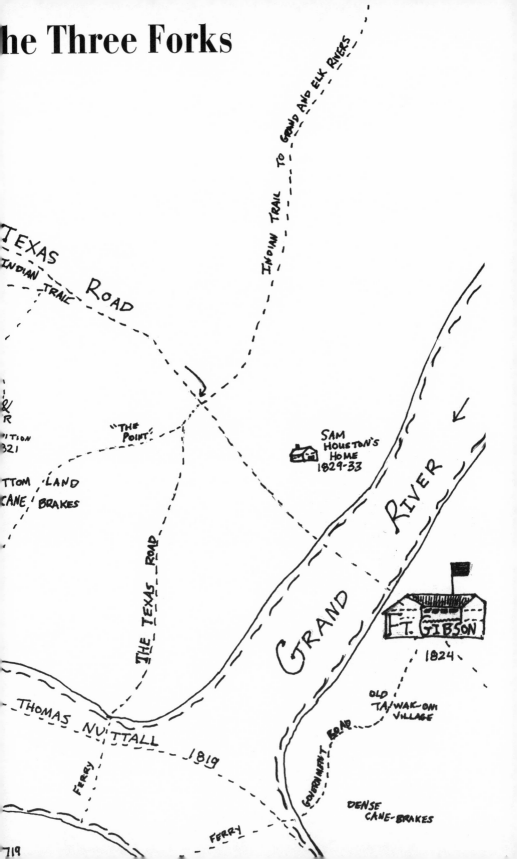

INDIAN TRAIL TO GRAND AND ELK RIVERS

TEXAS ROAD

INDIAN TRAIL

INDIAN TRAIL

&
R

ITION
321

TTOM LAND
CANE BRAKES

"THE POINT"

SAM
HOUSTON'S
HOME
1829-33

GRAND RIVER

THE TEXAS ROAD

FT. GIBSON

1824

OLD
TAWAKONI
VILLAGE

THOMAS NUTTALL

1819

FERRY

GOVERNMENT ROAD

DENSE
CANE-BRAKES

FERRY

Prologue

SEPTEMBER 9, 1819: Two men, weary travelers seeking the quickest way from the wilderness to civilization, food, and safety, arrived at the confluence of two great rivers. One, the Cimarron, they had been upon for a week; the other, the Arkansas, was broader, its water fresher, its current more compelling and dangerous. The exhausted men proposed to let this second river point the way, provide for their needs. It flowed through a land once part of a massive territory called Louisiana, currently part of a much smaller yet less amorphous territory called Arkansas. To these two weary travelers this land was as unknown, frightening, and dangerous as any hitherto seen.

The Arkansas River's course marked the past and the future. From its source deep in the Rocky Mountains the river flowed hundreds of miles, eventually reaching this place in time. Beginning as a violent torrent churning its way through walls of rock in the high country, the river eventually reached level, sandy terrain where its pace lessened and its girth widened. By the time it met the Cimarron, a tributary, the Arkansas was a great lazy river of immense sandbars and quicksand, shallows alternating with deep water as well as moving and stationary logs—sawyers and planters—creating a constant hazard to boatmen.

The two journeyers passed upon the river for a mere instant in time, a few days, before departing. But like other travelers before and since, thoughts of the river later dominated their memories. They recalled the sight of the hot sun glaring off the rippling water, the smell of evaporating moisture, the eerie drone of moving water, the deceptively passive appearance of the current that masks its hidden power of destruction and death. One of the travelers, Thomas

Prologue

Nuttall, had during the preceding fortnight grown familiar with images of death. A native Englishman, professional botanist, and Harvard College professor (in time), Nuttall was in comparison to his companion a greenhorn, foolish enough to drink from a tepid water hole during the hot, dry months on the prairie. He instantly became ill. It demanded tremendous willpower for Nuttall to survive, but he did so thanks in part to his resourceful companion, the otherwise unknown (and unchristened) Mr. Lee, a trapper who worked the waters of the Red, Canadian, and Cimarron Rivers; who purportedly had ascended the former almost to its source and the latter to near the Rockies; who himself had barely survived several run-ins with the native inhabitants; and who desperately wanted to rid himself of his two burdens, a load of beaver pelts that weighed down his canoe and an absentminded, foolishly courageous scientist who belonged in Philadelphia rather than in Indian Country.

The five weeks that joined the destinies of Thomas Nuttall and Mr. Lee occupy precious few pages of Thomas Nuttall's memoir *A Journal of Travels into the Arkansa Territory, During the Year 1819*. It was, however, because of their feuding interaction, reliance upon one another, conversations, and experiences of suffering and triumph that Nuttall was able to survive and provide an in-depth, elegant account of the wilderness prairie. The ways of Mr. Lee as well as those of other *natural men,* the Osage and Cherokee, the French boatman and American squatter, fascinated Nuttall, whose journal evokes and records a larger historical narrative of the dramatic and violent interaction of humans with the American wilderness. Nuttall's journal is a sophisticated narrative collage of natural history and human history in the trans-Mississippi West.

Nuttall reached the Arkansas with feelings of regret that his original goal, which had driven him from Philadelphia down the Ohio to the Mississippi, then up the Arkansas to Fort Smith, had not been accomplished. He had hoped to reach the origins of the Arkansas River in the Rocky Mountains. To this end he had hired the trapper Lee—whom he had met at Three Forks, at the confluence of the Arkansas with the Verdigris and Grand Rivers—to take him west

Prologue

across the prairie to the Cimarron River, the ascent of which would, Nuttall vaguely hoped, lead him to the foothills of the Rockies. But sickness and other misfortunes had driven the men east rather than west. Three Forks, rest and recovery rather than the Rockies, was Nuttall's only goal upon arriving at the confluence of the Cimarron and Arkansas. Thomas Nuttall, in short, was a stranger in a strange land who had come to the unknown wilderness voluntarily, to learn from it, but who would return to civilization terribly ill, worn out, beaten—yet wiser.

Thomas Nuttall was a great scientist, but he was an even greater explorer. Like other explorers he possessed an expansive imagination that conceived of the wonders of nature, placing himself vicariously among them, and at the same time offering the will and the stamina to make real his imagined experiences. Science and exploration were one in his mind, the former being a consequence of his wanderings, which were inspired by his joy at being a part of the natural environment, alone amid the grandeur of the creation. He had the gift of being able to analyze what he first intuited—he perceived beauty, then sought to understand it. In nature the bookish Nuttall was surrounded not by books but by an apparently infinite variety that revealed to him the wholeness of nature. This sense of unity gave him a wonderful sense of contentment. When alone in nature he was not lonely, and he sought solitude rather than the company of other humans.

The opening words of *A Journal of Travels into the Arkansa Territory, During the Year 1819* recall Nuttall's initial feelings of trepidation when he began his long journey: "I could not at this time divert from my mind the most serious reflections on the magnitude and danger of the journey which now lay before me, and which was, indeed, of very uncertain issue." Thought indeed precedes action, but in Nuttall's case a restless wonder preceded thought. One sees in his years of travel an image of the wilderness that was a counter to the experience of the present. The frontier lay at a distance—emotionally, figuratively, spatially, physically on the verge of civilization; there, away from conventions, customs, culture, and society, one

3

Prologue

could be embraced by the oneness of being. Nature was Nuttall's wife and children. It was an extension of self.

Nature was an extension of self for Mr. Lee as well, but in quite a different way than for his scientific counterpart. Lee did not approach nature from the choice of the imagination but from the necessity of the moment. Perhaps Lee was driven to the verge of civilization, the Missouri Territory, because of a criminal past, personal disaster, or utter poverty. Perhaps there was a more simple impulse behind his choice of vocation. There were many people such as Lee who went to the wilderness not because of flight but because of an urge from within, something like what compelled Thoreau to Walden Pond, up the Merrimack River, and into the woods of Maine—although without the philosophy and romance. John Filson's *The Discovery, Settlement and Present State of Kentucke,* published in 1784, which includes an appendage, "The Adventures of Col. Daniel Boon," characterizes this amorphous quality that is sometimes inexactly rendered by such words as "restlessnessness" and "wanderlust." Filson's Boone was a natural man, a common man become hero because of his frontier spirit. Filson quotes Boone as proclaiming that "it requires but little philosophy to make a man happy." Indeed, Boone was happy in solitude amid the whole of nature, which was imperfectly understood yet brought contentment. I suspect that Mr. Lee was the same. He fled civilization because of a compelling need to be with something that he could not find in civilization, could only find in nature. What was it? Did he find it? Thomas Nuttall doubtless believed that Lee's many wilderness experiences had allowed the trapper to discover this mysterious unnamed presence. Nuttall himself was only to find it on that September day in 1819 when he reached life's lowest ebb. It was then perhaps that he discovered the ability to accept himself, to rest content in movement and challenge, to take what is offered without asking for more.

1 Place of the Oaks, 1752

The Arkansas River, like all of America's great rivers, attracted for centuries Europeans and their descendants struggling to make real dreams of fame, riches, and enlightenment. The many rivers descending from the eastern slopes of the Rocky Mountains hosted a variety of wildlife, particularly the beaver, the pelts of which were prized by the wealthy of centuries past for their smooth durability and resistance to weather. Spanish and French traders bartered with the Indians of the Louisiana Territory for pelts, furs, and buffalo robes. The few hunters and traders who ascended the Red, Arkansas, Platte, and Missouri Rivers saw scores of buffalo reaching to the horizon and heard the thunderous noise of stampeding herds. Such plenty in a wilderness land brought forth images of undiscovered riches: cities of gold, mines of silver, crystal mountains, and unimpeded routes west to the Pacific. The source of these rivers was a mysterious place of fancy, a mirage in the sandy desert. Always in the distance were the great Shining Mountains where the Arkansas began, about which hunters, squatters, voyageurs, soldiers, and Indians talked at length, though rarely with firsthand knowledge of the river's glistening source.

This land of eastward-flowing rivers, varying climatic zones, contrary winds and conflicting air masses, violent storms, floods and drought, was and had been for years under contention. Once, centuries before, the Arkansas valley had been home to legend rather than fact. Spanish explorers, ever hungry for conquest and gold, imagined that the Aztecs and Incas were not the only New World peoples whom the Europeans could plunder. Stories of treasure to the north, at places called Cíbola and Quivira, drove greedy and restless Spaniards led by Francisco de Coronado across the Rio

5

The Land between the Rivers

Grande to the Red, Canadian, and Arkansas Rivers. Frustrated to find their energies wasted, their dreams just dreams, the conquerors left behind memories of random retaliation, senseless violence, and humiliating retreat. That same year, 1541, Hernando de Soto led his men up the Ouachita River into the mountains and river valleys that would in time be home to French and American hunters. De Soto accomplished murder and destruction, but little else.

One story has it that de Soto and his men wintered at a rendezvous spot for Indian hunters and traders, where ancient paths leading to and from rivers north and south crossed and met. Here was the confluence of three rivers—the Arkansas, Verdigris, and Grand—the Three Forks. Years before the Muskogean people, who called the Arkansas the Red River, had referred to the confluence of the three rivers as the "Forks of the Red River." De Soto himself heard the local tribes call his winter headquarters at the falls of the Verdigris River "Au-tiam-que." De Soto vanished with the spring— and so did the Spanish presence at Three Forks. Spain's vague claim to the lands north of the Red River had little actual substance. The regional Wichitas, Tawehash (Jumanos), Caddoans, Tawakoni (Touacara), Comanches, and Apaches continued to hunt and to war for scores of years. Meanwhile another European power, the French, came calling.

Robert Cavelier, Sieur de La Salle, having descended the Mississippi River to its mouth in 1682, and having seen along the way the mouths of other great rivers extending far to the west, claimed the land and its rivers for France, honoring his king with its name— Louisiana. Although many French explorers followed in La Salle's wake and ascended the rivers flowing from the west, the French never quite knew the extent of the land that they ostensibly governed. The Spanish, also ignorant of "Louisiana," countered French claims but did not have the energy or will to repel them. For several hundred years, in short, the Spanish and French maneuvered forces and trading posts in and about the Red River valley, from the Red River's confluence with the Mississippi northwest to the Great Raft,

Place of the Oaks, 1752

near what is today the Four Corners region of Oklahoma, Arkansas, Louisiana, and Texas. Here at the Great Bend the river makes a dramatic turn to the south. The Caddo and Wichita tribes farmed the rich valley of the lower Red and went on long hunts in the summer, trailing the massive buffalo herds northwest on the prairies. The dominant buffalo hunters of this land between the rivers were the Apache and Comanche tribes, fierce and more warlike than the Caddoans and Wichitas. The French, more apt than the Spanish or the English in treating the Indians as equals and understanding their culture, had friendly relations with the Indian tribes on the Red for several centuries.

One of the many voyageurs who journeyed inland, Bénard de La Harpe ascended the Red River to explore a grant of land given to him by the French governor of the Louisiana Territory. In 1719 he arrived at the Great Bend of the Red; established a log house and claim to the land; journeyed overland to the Canadian River; and then proceeded to the Arkansas, near where the river makes an abrupt turn northeast before proceeding in its general southeasterly direction (today southeastern Tulsa County). Here La Harpe traded with the Wichitas, partook in some of their ceremonies, and learned of the land and its varied peoples—including the Osage, Caddo, and Comanche tribes. The French were horrified to learn that the Wichitas were cannibals, feasting on war captives and slaves.

Two years later, downstream about twenty-five miles at Three Forks on the Arkansas, La Harpe met with thousands of Wichita, Tawakoni, and Tawehash warriors and elders. They smoked the calumet and traded. Burial mounds brought a somber air to the place. La Harpe pledged assistance to these Indians—nicknamed the "Three Canes People" in reference to the vegetation of the floodplain—with their enemies the Osage, Missouri, and Comanche tribes. He also promised more and better trade goods than those sold at Santa Fe. La Harpe was impressed with the rich and beautiful valley and declared the land part of the French-American empire. He departed, descending the Arkansas to its mouth and promising the return of the French to the Three Canes People, who

7

The Land between the Rivers

were, one assumes, less sanguine. The French, however, rarely capitalized on La Harpe's contacts. The random presence of French explorers such as Bourgmond and the Mallet brothers was of little importance in this world of repeated internal conflicts in and about the land between the rivers. For example, a successful invasion from the north by the Osages in 1752 forced the ouster of the Tawakoni and Tawehash, who retreated south of the Red River. The Osages called their new home "Place-of-the-Oaks." Here they rested after their long summer buffalo hunts.

The Osages were an aggressive, warlike people whose homeland was the mountainous region of the Ozarks in central Missouri. The Osages forged trails that traversed hundreds of miles of mountainous, forested, even arid western territory. One trail traversed Three Forks. As their influence extended west and south to the Arkansas valley, they established themselves as the most feared tribe of what is today Oklahoma. Like most of the Plains Indians, the Osage were seminomadic, relying heavily upon the meat, hide, bones, fat, and brains of the buffalo. They also grew legumes, pumpkins, and the like. They were a large, handsome people, made larger by legends of their ferocity and great strength. Men dominated in the tribe as hunters and warriors; women prepared food, made clothes from skin, and performed other tasks of domestic drudgery.

These newcomers to the confluence of the Arkansas, Verdigris, and Grand Rivers chose a wonderful site for their winter camp. The Arkansas, broad and shallow during winter months, its water brown and barely potable, received the clear waters of the Verdigris and then the Grand in short order. The Osages relied on these latter two rivers for drinking water. The area of Three Forks was well-watered, with numerous streams and heavier rainfall than found to the west. Oak, hickory, cottonwood, beech, sycamore, and mulberry trees grew in the floodplain and along the riverbanks, as did numerous bushes, brambles, cane, and grasses. Animals prized the region for the food and water. Hunters and trappers found not only buffalo but bear, deer, beaver, and rabbit. The rivers hosted bass, crappie, perch, bluegill, and other tasty fish. The Osages during these years

Place of the Oaks, 1752

lived generally good lives with scarcely any hunger. Provisions not otherwise provided at Three Forks were acquired by war, theft, or trade with other tribes, the Spanish, and the French.

Meanwhile war broke out in 1755 between the French and English. The French hoped to gain victories that would insure their dominance in North America. They lost the French-Indian War and with it their possessions in North America. New France (Canada) was surrendered to the English. But through sleight of hand the French passed control of Louisiana to the Spanish, who had wanted full control of it all along. While the French government vacated North America, the French-Canadians did not. They descended the Mississippi and ascended the Arkansas, and chose to stay in this no-man's-land to hunt and trade with the native tribes.

Whatever influence the French had upon local tribes such as the Osages largely ended with the passing of French Louisiana to Spanish control in 1763. The Spanish were less successful than the French in achieving friendly relations with the Caddoans, Wichitas, Comanches, and Osages. Spanish colonial authorities refused to invest the necessary resources to impose control over Louisiana, notwithstanding the efforts of explorers such as Pedro Vial. Almost forty years later the emperor Napoleon forced the Spanish to relinquish possession of Louisiana; soon after Napoleon sold the massive territory to the United States, which took control of a region that was still an unsettled land full of sporadic tribes competing for hunting grounds and warring with each other, as well as Spanish soldiers refusing to give in to the inevitable loss of the Red River and parts north.

The land between the Red and Arkansas Rivers continued to be for the Spanish, French, and English a place of dreams and legends and for the Native Americans a place to hunt, cultivate the soil, build shelters, and raise families. Varied tribes contended for the land and its plenty. In 1803 they were joined again by outsiders, this time Americans, who had bought the amorphous, still unknown Louisiana Territory.

2 Hot Springs, Winter 1804

The American acquisition of the vast Louisiana Territory from France in 1803 doubled the size of the United States and taunted American scientists with millions of unexplored acres west of the Mississippi River. The eastern scientific establishment, in particular the American Philosophical Society, responded with the Lewis and Clark expedition up the Missouri River and beyond the Rocky Mountains to the Columbia River and the Pacific Ocean from 1804 to 1806; the Dunbar-Hunter ascent of the Ouachita River in 1804; the Zebulon Pike expedition up the Mississippi River in 1805 and west to the Arkansas River and the Rockies in 1806; and varied other journeys up the Platte, Canadian, Cimarron, Red, and other rivers of the Louisiana Territory, across windblown prairies, and into the vast wilderness west of Lake Superior.

The United States, intent on securing claim to the Louisiana Territory, discovering its boundaries, exploiting its resources, and inserting a paternal control over its peoples, sent soldiers and explorers up the Missouri River in 1804. President Thomas Jefferson, representing the United States of America as well as the American Philosophical Society, recruited Meriwether Lewis and William Clark to discover the extent of Louisiana and the western mountains beyond. Legends suggested the existence of a water passage encompassing the whole of America, connecting the Atlantic to the Pacific. Lewis and Clark would reach the source of the Missouri in search of this Northwest Passage. The expedition followed in the path of French voyageurs, crossing the Continental Divide during the summer of 1805, descending the western slopes of the Rockies to the Columbia River, and reaching the Pacific Ocean at the end of 1805. Jefferson and other leaders of the American scientific com-

10

munity gave Lewis and Clark explicit instructions regarding their scientific activities and intended discoveries. Jefferson was particularly concerned that Lewis and Clark take almost daily compass readings and make frequent use of the sextant to determine location and latitude, from which an adequate two-dimensional picture of their path could be conceptualized by passive observers in the cities of the East Coast. To supplement such information the captains were to note depth of rivers, velocity of current, frequency of obstructions, and the extent and nature of the tributaries of the Missouri River. The president wished as well that the Indians of the Missouri valley and beyond, particularly the infamous Sioux and Blackfeet, know of the concern of the United States for those who were its friends and its antipathy toward those who were not. But more, Jefferson wanted clear information on the culture, customs, lifestyle, and beliefs of the Indians. And what were the natural products of the West available to Indians and Americans? Jefferson required Lewis and Clark to ship east, repeatedly and frequently, specimens of plants, animals, and Indian artifacts for perusal and study by members of the Philosophical Society.

The Lewis and Clark expedition put to rest many myths and established a basis for future exploration of and trade in the northern Louisiana Territory. Lewis and Clark discovered that the length and breadth of the Rocky (Shining) Mountains were much greater than hitherto suspected. They found by experience that hundreds of miles of rocky terrain separated the source of the Missouri from the source of the Columbia. They established relations with numerous Indian tribes that were more peaceful and sophisticated than Americans had been taught to believe. Meriwether Lewis and William Clark were intrepid adventurers, great explorers, and for the conditions under which they labored, fair scientists. Their collections were unique and their descriptions those of eyewitnesses. Yet some scientists who came to hear of the scientific exploits of Lewis and Clark wanted more precise information and were themselves eager to journey west in the wake of these pathfinders.

President Jefferson sent quite a different group of adventurers to

The Land between the Rivers

explore and acquire knowledge of the southern limits of the Louisiana Territory. In 1804 William Dunbar, a southern planter and slave owner, one of the best botanists of his time, and a member of the American Philosophical Society, responded to his friend Jefferson's request that he ascend the Red River to trace the extent and nature of the territory's southern border. Dunbar was not new to the rugged experiences of river travel, having descended the Mississippi River in 1773 to make his living as a planter. Like Jefferson he had broad scientific interests, ranging from astronomy to meteorology to geology. He was also an inventor.

The initial intent of his expedition, as revealed in the correspondence of Dunbar and Jefferson during the planning stages in early 1804, was to ascend the Arkansas River and return down the Red River. Dunbar, however, was worried about conflicts among the Osage Indians, who had split into two distinct bands, one living at the Verdigris River, one at the Grand River. Dunbar wondered whether or not he could obtain the services of a "guide & interpreter and hunter" who would be sufficiently familiar with both river valleys. By summer Dunbar had realized that such a journey would be impractical and so limited his proposal to the Red River. He was chiefly interested in discovering "plants useful for food in medecine [*sic*]" and the plentiful repositories of salt said to exist near the source of the river. Yet the explorer had to be wary of entering this region: "It is not improbable that the same mountains or chain of salt reigns along the ridge extending from the sources of the Missouri to those of the red river: the salt region is viewed by the savage native tribes as a sacred land never to be polluted by blood; the most inveterate Enemies asemble [*sic*] there for the purpose of collecting their provisions of salt, & so fully are they impressed with a religious & reverential awe of the Great Spirit supposed to preside more immediately over those parts that no example was ever known of public or private resentment having been satiated by those Savage men." Dunbar expressed to Jefferson during the summer of 1804 his desire also to explore the Ouachita River and the Hot

Hot Springs, Winter 1804

Springs, which had a "reputation for supposed Cures performed on some invalids who have bathed in [their] waters."

As William Dunbar made preparations for the journey, he formed a list of queries that he would pose to "old hunters in those Settlements [of the Red River] who have penetrated far into the interior & from [whom] useful & important intelligence may be obtained." Dunbar assumed such hunters would know the depth of rivers; the nature and quality of the soil; those plants "known to possess medicinal properties" and those seeds that could be collected; varied minerals found in the river valleys and hills; whether or not "Quadrupeds" such as the "mammoth" roamed the wilderness; whether or not the native tribes were "humane or vindictive"; "what are the accounts of gold, silver & other mines in the mountains"; and how to communicate with the Native Americans. Dunbar assumed that "a person well qualified to aid the party in the triple capacity of Interpreter, hunter and guide whose advice and opinions may be of the utmost importance to the Success of the Expedition" was "an indispensable expense; without such a provision" the expedition's members "might find themselves in the situation of a Ship at Sea without Compass or rudder."

Busy formulating plans during the summer of 1804, Dunbar received notice from President Jefferson of Spanish protests of American plans to explore the uncertain southern boundary of the Louisiana Territory; Dunbar changed his itinerary upon Jefferson's proposal that he journey up the Ouachita (Washita) River to the Hot Springs. A few months before the expedition began Dunbar informed Jefferson that "at the Washita Settlement we shall meet with old hunters and other persons capable of giving interesting details of the Countries high up the Western rivers; upon the whole I doubt not that we shall collect information sufficient to induce Congress to make liberal provisions for the important Expedition of the ensuing Season" to the Red River. Joining Dunbar was another scientist and "self-professed chemist," Dr. George Hunter. Dunbar and Hunter hoped to accomplish the modest objective of reaching

The Land between the Rivers

the source of the Ouachita River, a tributary of the Red. The Ouachita descended from a mountainous, fertile country that promised rich mineral deposits. Dunbar particularly wanted to investigate the warm waters of the Hot Springs, located near the source of the Ouachita, to confirm or deny the claim that they possessed healing attributes.

Jefferson provided Dunbar and Hunter with thirteen soldiers to row and pull their heavy keelboat upriver. The Dunbar-Hunter expedition lasted three months, from October 1804 to January 1805. The men ascended the sluggish and aptly named Red River to the clear, deep, and equally well-named Black River, which they followed to the Ouachita River. The region was moist and well-watered, dominated by cypress and willow trees. Wild fruits and honey trees abounded, as did plentiful supplies of venison and bear meat. Few Indians lived in the Ouachita valley. The river was named for the Washita tribe, since vanished. Some Choctaws, Cherokees, and Chickasaws had lately crossed the Mississippi to hunt this land between rivers. Otherwise the inhabitants were hunters, French traders, ne'er-do-wells, and a few honest squatters. Along the Black River the men refreshed themselves at "a small settlement commenced by a man and his wife: a covered frame of rough poles without walls serves for a house, and a Couple of acres of indian corn had been cultivated, which suffices to stock their little magazine with bread for the year; the forest supplies Venison, Bear, turkey &c, the river fowl and fish; the skins of the wild animals and an abundance of the finest honey being carried to market enables the new settler to supply himself largely with all other necessary articles; in a year or two he arrives at a state of independence, he purchases horses, cows & other domestic animals, perhaps a slave also who shares with him the labours and the productions of his field & of the adjoining forests." Such was Dunbar's happy assessment of the southern way of life.

A little way further they came to a river crossing where a French settler "keeps a ferry-boat for crossing men & horses traveling to or from Natchez and the settlements on [the] red river and on the

14

Hot Springs, Winter 1804

Washita river." The settler had received his grant from the Spanish a few years before; the land was extremely fertile, "equal to the best Mississippi bottoms," in Dunbar's opinion. The man was communicative, well-informed about the region. He listed the distance in leagues that Dunbar's men would have to row upriver to reach the furthest vanguard of civilization, the "Post on the Washita called Fort Miro." Having traveled thirty-two leagues from the mouth of the Red, and twenty-two from the mouth of the Black, the expedition now rowed into the Ouachita—with still nearly sixty leagues (180 miles) to go to reach the fort.

The Ouachita, like the Black, had a gentle current and was very shallow in places. There were signs of massive watery inundations in the past. Plants and trees that survive well in moist environments dominated the surrounding forest; there were numerous willows, magnolias, sycamores, even oaks of varied types. The oaks in particular had vines climbing and encircling their trunks and hanging from their branches like great wooden webs. The weather was sufficiently mild that some of the plants still flowered, even though it was near the end of October. The keelboat was large, with a draft deeper at times than the Ouachita. When the boat grounded the soldiers got into the cool water to tug at ropes and pry with poles.

Saturday, October 27, the boat was caught on a ledge just below the surface. It was a frosty morning. "A fog upon the river," Dunbar wrote in his journal, "occasioned by the condensation of vapor arising from the surface of the river," made the morning feel even colder. By afternoon they had advanced beyond the "embarass" causing the obstruction "into deep water" where "the river became again like a mill-pond without current, excepting a motion barely perceptible along the concave shore."

The last day of October 1804, the expedition arrived at another river crossing: "a ferry & a road of Communication between the Post of the Washita and the Natchez & a fork of this road passes on to the Settlement called the rapids on Red river, it is distant from this place by computation 150 miles." The locals informed Dunbar and Hunter that the scientists had selected the worst time of year to

15

The Land between the Rivers

ascend the Ouachita, as autumn was a season of very low water. The spring of the year was much better, in part because "the Mississippi then flows up into the beds of the inferior rivers, raising their waters sometimes within a few feet of the top of the banks." At such a time the keelboat would make more sense. The men daily found by experience how unwieldy their boat was. Twice on November 2 it was "fast upon a sunken log under water."

The autumn foliage was now at its pinnacle of beauty, the more so as the expedition ascended the river into the low hills of the Ouachita Mountains. Dunbar believed that the color of the leaves on a tree reflected the tint of the bark and wood when "extracted in the Dyer's vat." "I am persuaded," he wrote, "from the few observations I have made that this rule will be found general, and may therefore serve as an excellent guide to the Naturalist who directs his researches to the discovery of new objects for the use of the dyer." The trees grew in stature further up the river; along the shore "Cane began to appear, which is a sure indication of a fertile soil." Willows hung over the river, their thin, delicate leaves slowly changing from "their deep verdure" to "a fine deep yellow." Some days were foggy, even smoky; upon inquiring of the locals Dunbar learned that Indians and hunters set fire to the forest in the autumn to clear out the underbrush, which in spring would yield new growth of the "young tender grass" in which deer and other animals delight. Dunbar also learned from the locals that the Ouachita, which was clear and good to drink, was unlike the Arkansas, which was hardly potable: "The inconvenience from this . . . to voyagers, is not so great as might be apprehended, as it appears that brooks & springs of fine water falling into those rivers . . . are very frequent, and may be met often in the course of a days [*sic*] progress."

At the trading "post of Washita," near Fort Miro, the explorers were welcomed by Lieutenant Bowmar, his men, and the local French and American hunters and traders. Dunbar took time to make numerous observations of the sun and moon to indicate the latitude and longitude of the fort. The scientists exchanged their unwieldy and large keelboat for a "flat-bottomed barge," which

16

Hot Springs, Winter 1804

leaked so badly that she was "hauled ashore and caulked." The few inhabitants of the post engaged in the hunt during the winter and rudimentary planting in the summer. They paid exorbitant rates to the few merchants for basic supplies. Their diet consisted of meat and corn and of what little could be gathered from the forest. One of the locals "was hired at the rate of 30 dollars p^r month" to serve as the "pilot" who would guide them up the Ouachita River. This unnamed pilot became the source for much more information than directions. He told Dunbar that up the Red River, beyond the massive embarrass of trees and brush that blocked the ascent and descent of the river—the "Great Raft"—lay the territory of the Caddos, who were at war with the Choctaws and Osages. The Osage Indians were giants, the enemies of all, the hunter told Dunbar. Further along the expedition came to the encampment of an "old Dutch hunter." He claimed that he had been to the Arkansas River, that its wealth lay in an unending supply of salt as well as more limited but richer amounts of silver. The western mountains were enclosed by prairies, he informed the scientists.

Small creeks—bayous—continually fed the Ouachita, which was generally a passive river that was easy to ascend. As it neared the Hot Springs, however, the expedition journeyed through more mountainous territory, reflected in the more frequent rapids in the river. For successive days the soldiers and scientists struggled up one set of rapids after another. They poled and pulled the sluggish barge up "the Chutes" and "La Cascade," as local hunters called the most challenging impediments. On December 6 they landed and camped adjacent to a trail that led to the Hot Springs. The men set up camp at the springs, which issued from surrounding rock. The water was very hot, but palatable upon cooling. Dunbar found no evidence that it contained medicinal qualities, notwithstanding the claims of the soldiers and local hunters to the contrary.

The source of the Ouachita, near the Hot Springs, was at the eastern extreme of a range of small mountains that extended west for well over a hundred miles. The varied ranges of mountains in this region were largely unknown to men of science. They were small,

17

The Land between the Rivers

under three thousand feet, and forested all the way to their rounded summits, dominated by oak and hickory trees. On a clear evening looking west the hills appeared gray, even blue, with a background of pastel pinks and oranges painted by the setting sun. Few humans traversed the mountains; Indians generally kept to the larger rivers to the north and south. Nevertheless, deer, buffalo, even human traces wound through the woods, which were otherwise filled with thick bushes and vines. Grapevines encircled host trees, climbing a hundred feet to the treetops. The hanging vines gave an eerie look to the forest, particularly on cloudy days or at dusk. Local hunters had to be inured to silence and loneliness to spend very much time in such an environment of mystery. Legends and stories passed by word of mouth among the hunters of the Ouachita plateau, creating grand tales of crystal mountains gleaming with carbuncles or other treasure guarded by bloodthirsty panthers, wolves, and bears, as well as warlike Indians who might unexpectedly invade the haunted sanctuary of the forest. Locals told Dunbar about the distant Arkansas valley, inhabited by the Osage Indians, who terrified the few inhabitants of the region—but who fortunately rarely journeyed across the mountainous watershed that separated the Arkansas and Ouachita Rivers.

The return journey of Dunbar and Hunter down the Ouachita during the second week of January 1805 was rapid and easy. The weather was cold, though the Ouachita was reluctant to freeze. Dunbar continued his scientific observations and prepared an official report to submit to President Jefferson about the topography, geography, natural products, and people of the Ouachita valley. He included in the report what he had learned in conversations with two important early explorers: Colonel Francis Vego of Vicennes, Indiana Territory, and Auguste Chouteau, a French trader. Colonel Vego, he noted, "states that, in 1771 and 1772, he was engaged in the Indian trade on the Arkansas; that he ascended that river about 750 miles, and then entered and ascended one of its westerly branches." If Vego was correct in the estimate of his distances, he came close to reaching the source of the Arkansas in the Rocky

Hot Springs, Winter 1804

Mountains. Chouteau, one of the founders of St. Louis, was a sophisticated man of great and varied experiences. Dunbar could scarcely doubt his claim that he had ascended the Arkansas almost to its "head" and had at one point run a trading post "on the eastern declivity of the Mexican [Rocky] Mountains." He bragged that he was one of the few traders that the Osages allowed to live and trade among them.

Dunbar's report whetted Jefferson's appetite for more information. The president and scientist was quick to organize another scientific expedition to explore the region. In 1806 he sent Captain Thomas Sparks and scientist Peter Custis up the Red River. After a journey of seven hundred miles, during which they negotiated the massive and daunting Great Raft, they were stopped by Spanish soldiers, who questioned the American claim that the Red River was the southern boundary of the Louisiana Territory. The Spanish forced Sparks and Custis to turn back. The exploration of the Red River valley to near its source in the eastern foothills of the Rocky Mountains would await different explorers of a different time.

3 Three Forks, Christmas Day 1806

Captain Zebulon Pike, hundreds of miles away from his home and family, the memories and civilized comforts of the Ohio River valley, was on this winter day in 1806 camped next to the frozen Arkansas River in what is today Colorado. He and his two dozen men were hardly outfitted for a winter journey, even as they moved deeper into a mountainous territory known only to Spanish soldiers and Indian warriors. Pike, having ordered a Christmas Day rest, sat next to the campfire and wrote in his journal. He digressed from his typical habit of recording daily events and laconic observations; on such a day his thoughts turned to reflection on past joy, present suffering, and future uncertainty. "Here," he wrote, "800 miles from the frontiers of our country, in the most inclement season of the year—not one person clothed for the winter—many without blankets, having been obliged to cut them up for socks, etc., and now lying down at night on the snow or wet ground, one side burning whilst the other was pierced with the cold wind—such was in part the situation of the party." Meanwhile, several hundred miles downstream on the Arkansas River, Pike's second in command, Lieutenant James Wilkinson, and the seven men he led spent Christmas at the Osage Indian winter camp of Big Track, also known as Cashesegra. The camp was on the north banks of the Arkansas, about sixty miles away from the tribe's permanent village up the Verdigris River. Big Track declared his poverty even as he shared what little he had. Wilkinson and his men at least had something to eat, and the camp provided some protection from the wind. Compared to their comrades upriver, they had a cheerier Christmas.

Captain Pike was on his second expedition in as many years,

Three Forks, Chrismas Day 1806

exploring the boundaries and extent of the Louisiana Territory, recently purchased from the government of France by the United States. Thomas Jefferson, president of the United States (as well as of the American Philosophical Society), had sent Meriwether Lewis and William Clark up the Missouri River in 1804 and William Dunbar and George Hunter up the Red and Ouachita Rivers during the winter of 1804–5. Soon after, during the summer of 1805, Lieutenant Zebulon Pike had journeyed to fulfill another of Jefferson's goals: to discover the source of the Mississippi River. At the head of twenty soldiers and hunters, Pike ascended the Mississippi in a large keelboat, which eventually grew untenable as summer turned to autumn and the river began to freeze. George Fraser, Pike's pilot and a well-known hunter, directed the soldiers in replacing the keelboat with smaller dugout canoes, which the French called "pirogues," as well as durable and quick birch-bark canoes. Using such means of conveyance, Pike and his men endured incredible hardships pursuing objectives similar to those of Lewis and Clark, who were at the same time nearing the Pacific Ocean. Pike's goals were to make clear measurements using the compass and sextant, record observations on the landscape and peoples of the upper Mississippi, make rudimentary scientific analyses and collect natural specimens, and discover the source of the Mississippi River. Pike accomplished all of his goals save the last, and this he came close to accomplishing.

He was more successful in the expedition of 1806. This time Lieutenant (soon to be Captain) Pike set out from St. Louis across the central prairies of the Louisiana Territory. His orders, as presented to him by General James Wilkinson, were to befriend or intimidate the Plains Indians as the occasion warranted; explore the Arkansas and Red Rivers; and reconnoiter the southwestern extremes of the territory, which bordered the Spanish empire. His particular goal in the autumn of 1806 was to ascend the Arkansas River to its source in the Rocky Mountains, called by Pike and others the Mexican Mountains. Pike and his men reached the Great Bend of the Arkansas in what is today southern Kansas on October

The Land between the Rivers

18, 1806. The Arkansas was at that point so shallow as to be hardly navigable, with a twenty-foot-wide, passive stream coursing through a broad, generally dry bed. But a two-day rain changed that. The riverbed filled, though the water level continued shallow; Pike thought it sufficiently navigable. He split the expedition, sending Lieutenant Wilkinson, the son of Louisiana territorial governor General James Wilkinson, downriver, while Pike led men upriver seeking the Arkansas's source. Pike spent the first two weeks of November paralleling the descending Arkansas, marching west toward the mountains. The cold late-autumn winds impeded their progress. Cottonwoods, dwarfed by wind and sandy soil, dotted the riverbanks. The water was saline, and dried crystals of salt dotted the landscape. So, too, did sporadic prairie-dog towns, herds of wild horses, and even more herds of buffalo. On November 1 the men spied the distant peak later named for Pike, "which appeared like a small blue cloud." As they got closer the Rocky Mountains appeared white, "as if covered with snow, or white stone." The river ironically became more navigable as it narrowed. The water cleared with each day's travel. During the last week of November Pike attempted to ascend the "Grand Peak" but found himself deceived by the distances involved. So he contented himself as the Rocky Mountain winter set in to search for the origins of the great rivers of the Midwest. Through happenstance and error Pike ascended the Arkansas almost to its source.

The Arkansas, along with the Missouri, Platte, and Rio Grande, drains the cool, fresh waters of the eastern slopes of the Rockies. By the time the Arkansas has traveled the hundreds of miles to the Mississippi River it has grown to a massive bed with typically lazy, dirty water. Sandbars are its dominant phenomenon. Yet at its beginning, fourteen hundred miles away in the Rockies, the river has the energy of youth. Its water is green with cold and white with fury as it descends rapidly from the highest peaks, carving a watery niche through rock. At one point the river rushes through a narrow gorge with thousand-foot walls of bare stone.

Pike and his men stumbled upon this daunting feature—the

Three Forks, Chrismas Day 1806

Royal Gorge—in late December. The turbid rapids of the gorge were, astonishingly, frozen, which made for dangerous travel on slick ice dotted with jutting boulders, which were even more treacherous. Men and horses, cold and hungry, slipped and fell their way down the gorge, eventually arriving at snow-covered prairies in early January.

Lieutenant Wilkinson's journey was hardly easier. Wilkinson and his men departed on October 28 in a skin canoe big enough to hold five men and a dugout canoe holding two men, their equipment, and food. Pike thought "they appeared to sail very well"—at least for a hundred yards. Just out of their commander's sight the canoes "grounded" in the shallow water, which forced Wilkinson, the five soldiers under his command, and two Osage interpreters and guides "to drag" the canoes "through sand and ice five miles." Cold and snow soon froze the river, which made the canoes worthless; the men took to the shores. Travel was difficult. Lacking horses, the journeyers had to abandon most of their supplies. The desolate winter prairie—mostly treeless, entirely cheerless, gray and seemingly lifeless—disheartened the men, particularly Wilkinson, whose emotions naturally tended toward complaint and feelings of abandonment. The last week in November, the river rising, Wilkinson halted to construct dugout cottonwood canoes. The river refused to accommodate the voyagers, who met frequent shoals, shallows, and ice. At one point the canoes became icebound in midriver, which forced the men to wield their axes to cut a path through the ice. December 10 brought them to the confluence of the Arkansas and Grand Saline—the Cimarron. During the next two weeks the river depth was sufficient to allow their slow but unimpeded descent. On November 23 they halted at Big Track's camp near the Three Forks.

The Osage Indians who dominated the Arkansas valley and who made their homes, particularly during winter, at or near the Three Forks awed such Europeans and Americans who chanced to visit them. Few would seek out the Osages, the most feared warriors of the land between the rivers. Dunbar had found the hunters of the Ouachita to be terrified of the Osages and thankful that the water-

The Land between the Rivers

shed of mountains kept them in the Arkansas valley. Lieutenant Wilkinson and his party luckily escaped any conflict with the Osages, who were interested in Wilkinson's offer of a U.S. trading post, or factory, at the Three Forks. The lieutenant heard from hunters that Big Track was chief but only a nominal leader; the leading warrior was one Clermont, whose name meant "Builder of Towns."

French and English traders had for decades come to the Three Forks to purchase buffalo robes and horses from the Osages. The Osages were, however, usually suspicious of English traders, preferring the French, who knew their language and were willing to share in their customs. Frenchman Joseph Bogy (Bougie) ascended the Arkansas River to the Three Forks shortly after Wilkinson's reconnaissance. Bogy, energetic and fearless, at the same time knew how to adapt himself to such a place, where life mingled with the rushing waters of three rivers; where the demands of the wilderness engendered conflict; where chance became destiny in the lives of searching, questioning men, who learned nature's great lessons of surrender, humility, and acceptance.

4 Philadelphia, 1808

In the wake of the journeys of Meriwether Lewis, William Clark, William Dunbar, Zebulon Pike, and James Wilkinson, the Louisiana Territory was better known, yet overall still a land of mystery and uncertainty. Few knew this better than scientists such as Benjamin Smith Barton, a leader of the Philadelphia scientific community and the American Philosophical Society, nephew of the famous astronomer David Rittenhouse, and friend of Benjamin Rush, America's leading physician. Barton, one of the best botanists of his age, a professor at the University of Pennsylvania, and author of the first seminal study of natural life in America, *Elements of Botany* (1803), understood that knowledge would not come easily in the vastness of the Louisiana Territory, which extended north to the 49th parallel, south to the Red River, east to the Mississippi River, and west to the Rocky Mountains. Like his American Philosophical Society colleagues, Barton was a polymath, at home with a variety of topics of inquiry. He wrote well-regarded accounts on the origins of American Indians, studied Indian culture, researched American zoology, and was fascinated by the nascent study of the geology of North America. Barton, less the explorer and more the theorist, was constantly on the lookout for opportunities to gain further information to supplement what he knew about the Louisiana Territory.

Barton's students were some of the best botanical explorers of the young United States. Barton instructed Meriwether Lewis on the selection, identification, and preservation of floral specimens that Lewis would encounter on his journey up the Missouri River. The German immigrant Frederick Pursh became Barton's protégé in 1805. Pursh worked with Meriwether Lewis to catalog, sketch, and describe the flora of the Lewis and Clark expedition. Unfortu-

The Land between the Rivers

nately, the death of Lewis in 1809 intervened; Pursh, dissatisfied
with the project and his responsibilities, eventually moved to London, taking many of the plant specimens, notes, and drawings with
him. Another of Barton's students was Jacob Bigelow, author of
American Medical Botany (1817) and an explorer of note. Bigelow
ascended Mount Washington in New Hampshire in 1816 and wrote
a full account of the mountain's alpine flora in the *New-England
Journal of Medicine and Surgery*. Benjamin Smith Barton's most
famous student was Thomas Nuttall.

In 1808 the forty-two-year-old Barton chanced to meet a young
immigrant, a native of Yorkshire, England, who had just arrived in
America to satisfy an unnatural craving for knowledge of American
plants and to establish a reputation for himself in American and
European scientific circles as a careful practitioner of Carolus Linnaeus's taxonomic system. This young man, Thomas Nuttall, was
slight and good-looking, comfortable in polite society, possessing
the appropriate manners and moral assumptions of his time.
Although not an inheritor of the wealth and status of the English
gentry nor the beneficiary of a refined upbringing, Nuttall had edu-
cated himself to possess genteel manners and a dignified bearing.
Having been apprenticed to his uncle Jonas Nuttall, a Liverpool
printer, Thomas had stolen moments from his tasks to read and to
learn, to teach himself about the wonders of animal and vegetable
life. He had formed an acquaintance with the works of Linnaeus,
Buffon, and other naturalists; this acquaintance had grown into a
love for the incredible minutiae, the untold multiplicity of nature,
which had flowered into a passion for discovery. One imagines Nuttall bored with the tasks of the trade, unconcerned with the varia-
tions of business, daydreaming about another life, one unencumbered by shop hours and the demands of production. Daydreams
eventually generated a compulsion to journey to distant lands; an
obsession to trace the endless paths of wilderness; a will to endure
all trials, physical exhaustion, emotional pain, in order to achieve
the goal of observing, describing, collecting, preserving, and cate-

Philadelphia, 1808

gorizing delicate meadow flowers, the forest abloom with variety, land and waterfowl, wilderness fauna, and the native peoples of America.

Benjamin Smith Barton embraced the twenty-two-year-old Nuttall and began introducing him to American science and its great practitioners, men such as William Bartram, the naturalist and traveler, son of John Bartram, the greatest botanist of eighteenth-century America. Father and son had written seminal books on the natural history particularly of the middle and southern colonies and states. Nuttall studied John Bartram's *Observations on the Inhabitants, Climate, Soil, Rivers, Productions, Animals, and Other Matters Worthy of Notice* (1751), respecting the region from the Hudson River south to the Blue Ridge Mountains, as well as William Bartram's *Travels through North and South Carolina, Georgia, East and West Florida* (1791). One assumes that Nuttall peppered William, an old man in 1808, with questions. Barton convinced Nuttall to follow in the footsteps of the Bartrams, to devote his youthful energy to the arduous task of journeying to discover, catalog, and describe the unique and otherwise unknown variety of botanical specimens in America.

Under Benjamin Barton's tutelage Thomas Nuttall familiarized himself with American flora and the most important sources of information on American natural history, and gained a deeper understanding of the Linnaean system of taxonomy, which involved a two-word Latin phrase that described genus and species. Fortunately for Nuttall he was at the center of science in the United States at a vigorous time of intellectual activity, built on the equally vibrant scientific foundations of the past. From the beginning of the exploration and settlement of North America, naturalists had rigorously studied and left behind detailed accounts of the many unique plants and animals of America. Yet such was the scope of the object of inquiry and the incredible multiplicity and abundance of American natural history that the countless studies left much room for work from an ambitious and creative naturalist.

The Land between the Rivers

The beginnings had been made by the likes of the sixteenth-century Englishmen Thomas Hariot and John Frampton, scientists and writers on American materia medica; and the seventeenth-century Englishmen Lawrence Bohun, William Strachey, and John Josselyn, English naturalists and botanists who researched, traveled, experimented, and wrote on the vast cornucopia of American plants. Josselyn, for example, was an English physician who journeyed to New England on two occasions (in the 1630s and 1660s) to explore and botanize. He wrote two very influential books on natural history: *New-Englands Rarities Discovered* (1672) and *An Account of Two Voyages to New-England* (1674). Eighteenth-century scientists who provided more information on American natural history included John Lawson, an English explorer who lost his life to Indians; Mark Catesby, who published a beautifully illustrated *Natural History of Carolina, Florida, and the Bahama Islands* in 1732; and Johannes Gronovius, who authored *Flora Virginia* (1739), based on the collections of the exploring botanist John Clayton. Clayton was friend to the English botanist John Ray, as well as the Virginia naturalist William Byrd II. Byrd was a member of the Royal Society of London, which after its formation in 1662 provided continual stimulation for colonial botanists. Also providing significant inspiration was the great Swedish scientist Carolus Linnaeus, who had many friends and correspondents in America, such as John Mitchill, a botanist and a member of the Royal Society. Linnaeus sent his protégé Peter Kalm to America for botanical information in 1748, which inspired Kalm to publish his *Travels in North America* (1770). James Logan of Philadelphia was another patron of botanical research, as well as a scientist of note, who left his large scientific library to the city of Philadelphia upon his death.

During the latter half of the eighteenth century and the first few years of the nineteenth, the work of botanists such as the Bartrams and another father-son team, the Michauxs, expanded the range of botanical activities to the southern colonies and states and across the Appalachians to the trans-Appalachian forests of the Ohio and Mississippi river valleys. The *Transactions* of the American Philo-

Philadelphia, 1808

sophical Society, which Nuttall could easily consult, provided numerous botanical papers and travelers' accounts. Manasseh Cutler, for example, a Massachusetts botanist, ascended one of the highest peaks in the Appalachian Mountains to record the amazing flora of Mount Washington. The *Transactions* included other important botanical records such as Gotthilf Muehlenberg's accounts of Pennsylvania flora. And then, of course, there was Barton's *Elements of Botany* to master.

Philadelphia was a busy city, the cultural and scientific capital of the young United States. It had been home to the versatile Ben Franklin, who continued to inspire Philadelphia citizens to diverse intellectual achievements. Although the great scientific leaders of the eighteenth century had passed on—James Logan, John Bartram, David Rittenhouse—others of the nineteenth century continued their erudite legacy—William Bartram, Benjamin Smith Barton, and Benjamin Rush. Rush was an old man in 1808, yet still at the top of the medical field in America. His reputation as a student of medicine, disease, and chemistry was soon to be accentuated by his treatise *Diseases of the Mind* (1812). Rush was professor of medicine at the University of Pennsylvania Medical College; upon Rush's death in 1813 Barton replaced him as professor of medicine. Other Philadelphia scientists included the geologist William Maclure; the anatomist Caspar Wistar; the ornithologist Alexander Wilson; and the engineer Erskine Hazard, son of the polymath Ebenezer Hazard, one of Philadelphia's intellectual lights in the late 1700s. The city boasted wonderful engineering accomplishments during the first few decades of the 1800s: a turnpike to Lancaster had recently been opened, as had a new bridge over the Delaware River, under which the new phenomenon of the sea, steamships, entered the harbor of the second-largest city in America. Until recently (1800) Philadelphia had been the political capital of the United States while Washington, D.C., was being designed and constructed. Philadelphia continued to be a leader in America's economy: the second Bank of the United States opened there in 1817. It was also the center of America's emerging insurance industry. But most

The Land between the Rivers

important, of course, was that Philadelphia towered above other cities in the sciences. Besides the venerable American Philosophical Society, there existed the Academy of Natural Sciences, the Philadelphia Medical Institute, and the *Philadelphia Journal of Physical and Medical Sciences.* Such were the reasons that attracted Thomas Nuttall to Philadelphia in 1808.

Professor Barton sponsored Nuttall on several expeditions in search of plant specimens. The first was to coastal Delaware in June of 1809, when Nuttall got his feet wet in exploring the American wilderness, in this case the swamps of Delaware. Nuttall was introduced as well to the necessity of relying on others, particularly local farmers, hunters, and trappers, to serve as guides into the wilderness. Such a wilderness was the Great Cypress Swamp in southeastern Delaware, which Nuttall dared to traverse only with the help of "an old man who usually conducts strangers into the swamp." The swamp was the first of many "frightfull labyrinths" that Nuttall would traverse over the course of his many journeys: "It was filled with tall tangling shrubs thickly matted together almost impervious to the light." The mosquitoes so bit Nuttall that his face appeared infected with viral corpuscles.

In August Nuttall journeyed for Barton northwest to the Pocono Mountains and the beautiful hamlet of Wilkes-Barre, nestled in the Susquehanna valley. Traveling the same route as had John Bartram in 1738, Nuttall paralleled the Susquehanna north to New York, at which point he turned west to the lake regions, Buffalo, and Niagara Falls, which he proclaimed "the most astonishing production of nature." After exploring the environs of Lake Erie, Nuttall crossed the Niagara River into Ontario, then journeyed along the coast of Lake Ontario to Ancaster, where he spent a frustrating month recovering from a debilitating attack of the ague. Fever, chills, nausea, and vomiting characterize the ague, which is a consequence of malaria. Nuttall would repeatedly experience these symptoms of malaria during subsequent years, particularly at the most inopportune moments.

5 Detroit, Michigan Territory, Summer 1810

The initial journeys of Thomas Nuttall and his sufferings consequent upon the pursuit of botanical knowledge in the American wilderness did not dampen the young scientist's enthusiasm; rather, they encouraged him to journey further west in search of the botanical unknown in particular and the unknown of natural history in general. One gets the impression that by 1810 Nuttall had set his mind on exploring the whole of America from the Appalachians west to the Rocky Mountains. To this end he journeyed in 1810–11 into the Indiana and Michigan Territories and up the Missouri River to the Yellowstone River, then down the Mississippi River to New Orleans.

Ostensibly Nuttall was Barton's employee. Barton had an agreement with William Clark to produce a complete natural history of the Lewis and Clark expedition. Nuttall was to journey into the general area first traversed by Lewis and Clark, collecting specimens and making copious journal entries. He was to be paid eight dollars a month. Barton, having been disappointed in his relationship with Frederick Pursh, was confident that Nuttall would bring about the success of his botanical journey. Barton described Nuttall at this time as "a young man, a native of England, brought up in a manner under my own eyes and instruction, and distinguished by his love of science, his integrity, his sobriety, and *innocence* of character."

Thomas Nuttall departed in April 1810 on this second journey into the interior United States with the confidence of youth, oblivious to the numerous scrapes he would find himself in, the narrow escapes from death he would experience. Barton charged Nuttall

The Land between the Rivers

with a host of unrealistic, Herculean tasks that could scarcely be accomplished by one man with little knowledge of woodcraft and self-defense traveling alone through wilderness regions. Barton's enthusiasm told him that Nuttall could journey across Pennsylvania, north to Detroit, south to Chicago, north along the western shores of Lake Michigan to Green Bay, then on to Lake Superior, from which the pilgrim should proceed northwest into the Canadian wilderness to 55 degrees latitude—"beyond which," Barton instructed, "I do not wish you to proceed, unless you shall find, from correct information, strong inducements to do so." Nuttall fortunately had the common sense on the journey to ignore most of Barton's instructions and follow his own interests and instincts. He must have in time reflected on the absurd understatement with which Barton concluded his instructions: "Always remember, that, next to your personal safety, science, and not mere conveniency in travelling, is the great object of the journey. In pursuit of curious or important objects, it will often be necessary to court difficulties. . . ."

Nuttall discovered the extent of the difficulties rather quickly. The initial fatigue of travel brought on a recurrence of the ague, with which he would struggle off and on for the remainder of this two-year-long excursion.

Nuttall began his journey by taking the stagecoach to Pittsburgh, departing April 12. Trees had not yet blossomed, and flowers generally lay "in dormant sleep." Songs of spring were nevertheless heard from the eastern bluebird (*Sialia sialis*), "the first agreeable presage of returning spring," his bright blue coat distinct in the still gray branches of late-blooming trees. Nuttall thought the bluebird acted and sounded like the mockingbird (*Mimus polyglottos*), though the latter's song was much superior to the bluebird's. Nuttall observed as well the horned lark (*Eremophila alpestris*), the reclusive tree swallow (*Tachycineta bicolor*), and the red-winged blackbird (*Agelaius phoeniceus*), beautiful "harbingers of approaching spring." Another important sign of spring was the blossoming of the great laurel (*Rhododendron maximum*), its beautiful pinkish-white flowers beginning to show in the valleys of the Alleghenies. At Bedford Nut-

Detroit, Michigan Territory, Summer 1810

tall walked to the nearby mineral springs, from which, in anticipation of the ague, he "drank pretty freely of the water expecting another fit the next day particularly as I had neglected taking my usual medicine." Although Nuttall also found the trailing arbutus, or mayflower (*Epigaea repens*), abloom, pausing to smell its delicious scent, there is no indication that he made a tea of its leaves, as would native tribes for the nausea that sometimes accompanies the ague. Also good for such illness is sassafras (*Sassafras albidum*) and the windflower, or rue anemone (*Thalictrum thalictroides*), its white flowers beautiful, which Nuttall spied on Laurel Mountain. Along the way Nuttall met a fellow passenger, Manuel Lisa, the great and notorious explorer and trader of the Missouri River. Lisa's verbal narratives set Nuttall to thinking about the possibilities of botanizing on the Missouri River. Perhaps Lisa was with Nuttall when the latter spied a ruffed grouse (*Bonasa umbellus*) making threatening gestures.

At Pittsburgh Lisa and Nuttall parted; Nuttall remained in the town for several days. He toured the growing city with a Dr. Stevenson. On Sunday, April 22, Nuttall explored the banks of the Monongahela River, ascending along the shore about five miles. The Monongahela descends from the south and merges with the Ohio at Pittsburgh. On his walk Nuttall saw the toothwort (*Cardamine diphylla*), which the locals used to ward off toothache, headache, or any other discomfort. He spent the night at a tavern, at which he learned from the proprietor of small alligators (*Salamandra horrida*) sometimes caught in the river. During the night he could hear the cry of the whippoorwill (*Caprimulgus vociferus*) coming from within the forest; some locals considered its distant, eerie call "an omen of misfortune." Nuttall, fascinated by its "confused vociferation," had never heard the whippoorwill before his arrival in America.

On Wednesday, April 25, Nuttall turned north, paralleling the beautiful Allegheny River on its path through rolling, wooded hills. The fever, chills, and vomiting of malaria recurrently interrupted his singular aim to examine all remarkable phenomena: salt springs and petroleum deposits; the common illnesses of the region, such

33

The Land between the Rivers

as goiter; the numerous wolves; the wonderful variety of birds. The latter particularly fascinated Nuttall, as his ornithological interests were second only to his botanical. Indeed, he would in time publish a two-volume *Manual of the Ornithology of the United States* (1832, 1834).

The second day out from Pittsburgh was hot; Nuttall became fatigued and thirsty and stopped at a "log-house" tavern for refreshment. He immediately became ill but "happened to meet with kindness & attention from these people." He was well enough the following day to make an eight-mile excursion to examine a saltworks, salt being one of the most valuable productions of western Pennsylvania. He heard from the inhabitants of Butler that the goiter was growing less prevalent, though many were troubled with the "Billious" fever and rheumatism. Fortunately honeysuckle (*Lonicera*) and dwarf ginseng (*Panax trifolius*) grew in the region to treat rheumatism; the nectar of the former was a tasty treat, while the root of the latter could stave off hunger. Nuttall's problem, the ague rather than rheumatism, bothered him for several days at the end of April. He tried an "emetic" that became "cathartic"—the combination of which left him very weak. On the evening of May 4 Nuttall spied a striking belted kingfisher (*Ceryle alcyon*) skimming the waters of the Allegheny in search of fish. After another spell of sickness at Franklin he continued paralleling the Allegheny to a local "*Petroleum* springs." The high banks and rough trail along the Allegheny tired him sufficiently that he stopped at a local house for the night. Amid his fatigue the beautiful scarlet of the fire pink (*Silene virginica*) and the deep blue of the vetch (*Vicia americana*) served as a cordial. The hospitality of the Haliday family helped as well; Nuttall was forced to spend the next day (May 9) recovering from the ague. He could not shake the fever, chills, and nausea for several days and so made slow progress toward his destination of Le Boeuf.

Along the way he "heard the gobling [*sic*] of the wild turkey" (*Meleagris gallopavo*), a bird he was most interested in. Nuttall observed turkeys on most of his travels west of the Appalachians. He found their odd behavior and look fascinating. They rarely use their

34

Detroit, Michigan Territory, Summer 1810

wings for flight, except when pursued by a predator that they cannot outrun; "their speed," he wrote, "is very considerable." They are wont to roost in trees, betraying their location "at dawn" with "a *gobbling* noise." Nuttall was not, however, astonished by the turkey's intelligence: "When approached by moonlight," he wrote in his *Ornithology*, "they are readily shot from their roosting-tree, one after another, without any apprehension of their danger, though they would dodge or fly instantly at the sight of the Owl. The gobblers, during the season of their amorous excitement, have been known even to strut over their dead companions while on the ground, instead of seeking their own safety by flight."

The arbored environment in which Nuttall saw the turkey was a swampy land of the immensely tall white pine (*Pinus strobus*); the verdure hemlock (*Tsuga canadensis*); the goldthread (*Coptis trifolia*), employed by the Indians and settlers for canker sores; the pink and white painted trillium (*Trillium undulatum*); the black currant (*Ribes americanum*), useful for ridding the intestines of worms; the bluebead lily (*Clintonia borealis*), useful on inflammations of all kinds; the Canada violet (*Viola canadensis*); and the blue-eyed Mary (*Collinsia verna*), which Nuttall named in honor of his friend and patron Zaccheus Collins of Philadelphia, who was an accomplished botanist, member of the American Philosophical Society, and a warm supporter of Nuttall's work. The sweet song of the rose-breasted grosbeak (*Pheucticus ludovicianus*) accompanied the lovely profusion of wildflowers.

Unhappily Nuttall traveled with a trunk, concern for which cost him much time and effort. He dragged it on the journey along the wilderness trail from Pittsburgh paralleling the Allegheny, his body burning with fever. After determining to leave the trunk behind at Franklin to be brought forward by boat while he continued on his pedestrian way, he discovered, having struggled through swamps to reach Le Boeuf, that the boat's itinerary was not what he had been led to believe; its route through French Creek was delayed. The trunk remained at Franklin. Disheartened and ill, Nuttall decided to return to Franklin. It was sufficiently difficult to travel with gun

The Land between the Rivers

and powder, pen and notebooks, thermometer and scales, much less a trunk! Thomas Nuttall would learn by trial and error what worked and what did not, in preparation for greater and more arduous journeys.

During the last two weeks of May Nuttall struggled through swampland, retracing his path to Franklin. For several days he traveled "with a heavy heart brooding on my disastrous journey." Nature provided diversions, and even cordials, to bring Nuttall's spirit back to its normal exuberance. The yellow lady's slipper (*Cypripedium pubescens*), a showy, delicate yellow, was used by Nuttall's contemporaries for depression. Once Nuttall reached Franklin, however, he was stricken with "the severest attack of the ague that I have ever yet experienced. I feel extremely discouraged by disappointments & this never ending disease. I should gladly have relinquished this disastrous journey had I not found agents giving healing & fortitude & known to administer it when it became absolutely impossible to proceed any farther." At Franklin he bled himself and "tried Calomel & jalap, which acted as a violent emetic, which I stood much in need of." At other times Nuttall used tea made of the leaves of thoroughwort, which the colonists called "boneset" (*Eupatorium perfoliatum*), to reduce the severity of his malaria. American settlers had over the years learned from the Native American materia medica of many other plants to combat the symptoms of the ague. Nuttall was familiar with herb Robert (*Geranium robertianum*), a geranium that made a good tea for a variety of ailments, including the ague. Nuttall found the flowering dogwood (*Cornus florida*) in the Alleghenies; Americans knew that tea made from its bark was a good substitute for quinine. The cucumber magnolia (*Magnolia acuminata*) also made a good bark tea for the ague; Nuttall discovered this beautiful flowering tree along Lake Erie.

Returning to the trail again on his second journey to Le Boeuf, Nuttall carried his trunk and gun as he struggled through the wilderness. It took five days to reach Le Boeuf, two of which he

Detroit, Michigan Territory, Summer 1810

spent with a "hospitable farmer" and another frontiersman who lived at "the entrance of 16 miles of wilderness." Nuttall was not too sick to make notes of his discoveries, such as the milkweed (*Asclepias quadrifolia*), abloom with white flowers, and the bluebead lily (*Clintonia borealis*), its berries just emerging, which he had seen on his first journey to Le Boeuf. The day of his arrival at Erie, Pennsylvania, Nuttall saw a wild turkey (*Meleagris gallopavo*), a short-tailed shrew (*Blarina brevicauda*), and a red-tailed hawk (*Buteo jamaicensis*), no doubt interested in the shrew as well. It is not unusual to see any of these three in the woods, in particular the hawk, its shrill call often breaking the silence of the forest, its beautiful white and black underside visible as it soars high above on bright sunny days. Nuttall could not think of a rapacious hunter equal to this hawk in search of food on winter days.

At Erie, frustrated at getting a passage by boat and impatient to be off now that he had enjoyed a few days of health, Nuttall continued his pedestrian journey, using the great lake for his bearings, traveling west with Lake Erie to his right. As he hiked the beach these early days in June 1810, the leaves of trees were young and fresh, a bright healthy verdure. He saw the cottonwood (*Populus deltoides*), its leaves delicately fluttering in the breeze. The zephyrs of June 2 became "2 different tornadoes" on June 3; heavy rains and violent winds crashed the trees about and "drenched" the scientist, who took refuge in a semidry "hovel." The hovel became his home for a day because of a return of the ague. Fortunately the cucumber magnolia (*Magnolia acuminata*) stood nearby. Stands of beech trees (*Fagus grandifolia*) provided excellent shade. Matching the gray bark of the beech was the gray skin of the skink (*Eumeces fasciatus*). Other reptilian creatures greeted Nuttall upon his entry into Ohio and arrival at the Grand River, where the young settlement of Painesville stood. The garter snake (*Thamnophis sirtalis*) was ubiquitous, the timber rattlesnake (*Crotalus horridus*) fortunately not as much. Both snakes gloried in the forest of hickory (*Carya*), oak (*Quercus*), birch (*Betula lenta*), buckeye (*Aesculus glabra*), maple

The Land between the Rivers

(*Acer rubrum*), and ash (*Fraxinus americana*). Numerous chipmunks (*Tamias striatus*) ran about the roots and forest undergrowth while gray squirrels (*Sciurus carolinensis*) chattered at all who came near.

Cleveland, at the mouth of the Cuyahoga River, on the southern shores of Lake Erie, was when Thomas Nuttall visited in June 1810 an insignificant hamlet. Rain forced his delay at Cleveland for several days, which he spent observing and collecting. He collected the roots and flower of the loosestrife (*Lysimachia quadrifolia*), the former of which in a tea was a good emetic. This wildflower grows well in swamps, which describes the environs of Cleveland. Also growing in wet environments is devil's bit (*Chamaelirium luteum*), good for intestinal complaints.

A boat for Detroit failing to appear, Nuttall proceeded happily along the shores of Lake Erie, which on the evening of June 13 "was serene & calm; scarce a zephyr disturbed the composure of its waters which spread themselves far & wide like a vast mirror, but the calm of its surface was strongly contrasted by the rugged aspects of its dangerous cliffs of its banks which are high perpendicular & sometimes dreadfully projecting rocks advancing abruptly into the water. These banks continue for many miles and have often proved fatal to the unsheltered marriner [*sic*] in the storms to which these Lakes are very subject."

On June 14 Nuttall made several discoveries. He found wild garlic (*Allium canadense*), which the Indians used to treat a long list of ailments. The beach of Lake Erie abounded with snakes. The inland woods included such fowl as the wild turkey, which Nuttall closely examined and declared "considerably tame," and the snowy owl (*Nyctea scandiaca*), which he did not see but assumed was what the locals meant by "a monstrous white *owl*." The woods included as well a storehouse for the apothecary: Solomon's seal (*Polygonatum biflorum*), false Solomon's seal (*Smilacina racemosa*), the Canada mayflower (*Maianthemum canadense*), and valerian (*Valeriana officinalis*) helped Indians and settlers fight anxiety, depression, intestinal problems, coughs, and so on.

Near Ohio's Huron River, flowing north, he walked through an extensive prairie, which gave him a taste for his future travels along the Arkansas River. Nuttall saw the bobolink (*Dolichonyx oryzivorus*) amid the tall grasses, feasting on the seeds and insects common to prairies. The festive nature of its song sounded like "Bob-o-link, Bob-o-link, Tom Denny Tom Denny. Come pay me the two and six pence you've owed more than a year and a half ago!"—such, at least, was how the "boys of this part of New England," Massachusetts, rendered it, according to Nuttall's *Ornithology*. "However puerile this odd phrase may appear," he added, "it is quite amusing to find how near it approaches to the time, and expression of the notes, when pronounced in a hurried manner. It would be unwise in the naturalist to hold in contempt any thing, however trifling, which might tend to elucidate the simple truth of nature. I therefore give the thing as I find it."

This region—once part of the Northwest Territory, then the Indiana Territory, before becoming in 1804 the Michigan Territory—was well-watered by streams and rivers. These were rapid, with little available portage. There was a profusion of swamps as well to frustrate the designs of the pedestrian traveler. Yet the colors of early summer distracted Nuttall from irritations of the journey. The forested landscape had become a deep verdure, dominated by deciduous hardwoods such as the stately sycamore (*Platanus occidentalis*) and the black walnut (*Juglans nigra*). The forest featured as well beautiful elms (*Ulmus*), maples (*Acer*), and a host of coniferous trees. Wildflowers that caught Nuttall's attention included a species of milkweed (*Asclepias*), which Indians used for a variety of ills ranging from constipation to asthma.

Upon reaching the mouth of the Huron, he took passage on a boat west and north on Lake Erie to the mouth of the Detroit River. Up the river at a graceful bend was the small, growing town of Detroit. Nuttall stayed at Detroit for over a month, exploring the environs. The land was fertile and held great potential for the New England immigrants arriving daily. Although Detroit had a reputation for being especially healthy, its inhabitants commonly suffered from goiter, an illness of the thyroid that involves swelling of the

The Land between the Rivers

neck, caused by insufficient iodine (from salt) in the diet. Because of the lack of other freshwater sources, the inhabitants relied on the river for drinking water. Some thought that the river water was to blame for the goiter, but Nuttall declared it to be "very pure," adding that "the water of the lakes is perhaps as pure & generally wholesome as any body of fresh water in the world."

Detroit was an old French fort, originally called Pontchartrain by its founder, Antoine de la Mothe Cadillac. The fort had passed to British control in 1760 during the French-Indian War, when Major Robert Rogers accepted the French surrender. The native tribes of the Detroit River, led by the Ottawa chief Pontiac, were suspicious of British intentions and frustrated at British arrogance. Pontiac's Rebellion was the result, during which Fort Detroit was under siege by but never fell to the native attackers. Detroit was a British fort until the end of the War for American Independence, when it became part of the new Northwest Territory. By the time of Nuttall's visit Detroit was capital of the newly formed Michigan Territory. Though the settlement had changed hands three times in fifty years, its inhabitants were largely French, and the place names of the region continued to be of French and Indian origin. The dominant natives of the region were the Chippeways and the Hurons. The Hurons were superstitious, believing in witches, which if caught were put to death. Also superstitious were the French Catholics of Detroit, at least according to Nuttall, a communicant of the Church of England. When Nuttall saw Detroit it was a pleasant-looking town of fifteen hundred situated on a bluff above the river, paths from the many wharves leading up to the white one- and two-story houses of the inhabitants. It had been only five years since Detroit was destroyed by fire; the town had made a remarkable recovery.

During Thomas Nuttall's month-long stay at Detroit, he spent his days exploring, asking questions of the local inhabitants, and collecting data and specimens. He was particularly intrigued with the star-nosed mole (*Condylura cristata*), the deer mouse (*Peromyscus maniculatus*), the gar (*Lepisosteus longirostris*), and the catfish (*Ictalurus punctatus*). The purple martin (*Progne subis*) was a worthwhile

Detroit, Michigan Territory, Summer 1810

bird to know, courageous in driving away hawks and crows and willing to associate with humans on any level, equally pleased "with the master and the slave, the colonist and the aboriginal. To him it is indifferent, whether his mansion be carved and painted, or humbled into the hospitable shell of the calabash or gourd." Another bird that Nuttall saw in the environs of Detroit was the ruby-throated hummingbird (*Archilochus colubris*), which so fascinated him that in a later year, domiciled on the East Coast, he kept one as a pet. Nuttall also observed now extinct passenger pigeons (*Ectopistes migratorius*). These birds fascinated Nuttall with their incredibly great numbers, which made their movements appear like a great black cloud presaging a storm. "The whole air is filled with them," he wrote in *Ornithology;* "their muting resembles a shower of sleet, and they shut out the light as if it were an eclipse."

Trees that Nuttall observed included those common to the area, such as the white ash (*Fraxinus americana*), the American beech (*Fagus grandifolia*), the American basswood (*Tilia americana*), the bee tree of the frontier, and the American plum (*Prunus americana*). Of especial interest was the paper birch, or white birch (*Betula papyrifera*), a distinctive white-barked tree of the northern forest. Nuttall explained in his journal the efficacy of the white birch, "which is extensively used by the Indians for constructing canoes about which they display considerable ingenuity & skill. These boats are much used by the fur-traders being very light of carriage across the numerous portages of the N. West & Lake countries & also admirably adapted for navigating the shallowest streams." All of this Nuttall found out in the ensuing months of his journey.

In late July 1810 Nuttall contrived to join the surveyor of the Michigan Territory, Aaron Greeley, on his journey to Mackinac (Michilimackinac) Island. In a large birch-bark canoe, which could hold close to a dozen men as well as several tons of supplies, they ascended the Detroit River the few miles to Lake St. Clair, after which the St. Clair River led them to Lake Huron. Lake Huron, like Lake Erie, recalled the original peoples of the region, the Hurons and the Eries. Nuttall sought information on the Hurons as well as other tribes still in existence, the Chippeway, Delaware, and Iroquois. Nuttall had a chance to interview a Chippeway native, who informed the scientist of their material culture and weaponry. Nuttall was sufficiently sensitive to realize that the information he collected was the product of oral tradition, the veracity of which required a different means of assessment. "With the Americans as with other people," he confided to his journal, "traditions are not known to all, neither are they at the same time committed to any particular sect, but all are not historians, all are not warriors, neither are all orators; & while one man listens with anxious curiosity to the tale of ancient customs, another passes them over unheeded, or recollects them indifferently, & neglects to transmit them to his posterity, so it is perhaps with the Indians."

The native tribes of the region, who had at some point in time erected fortifications, impressed Nuttall with their knowledge; engineering skill; choice of defensible, fertile, well-watered locations for their towns; and the sophistication of their material productions. He wrote, for example: "I have seen pieces of embroidery with Porcupine quills which both for taste & neatness of workmanship would scarcely be excelled in Europe with the same means." Yet the Indians had not yet broken from the unending cycle of war and retaliation; hence their population remained low, their foodstuffs scanty, and their culture limited. From his stay at Detroit and subsequent journey north Nuttall could date his growing interest in discovering and recording the traditions of the native Americans.

During the voyage of several weeks Nuttall got his first real taste of the lives of French boatmen—voyageurs—their gruff cheerfulness, their fears and superstitions, the monotony yet fascination of their lives. Every scene was new to Nuttall. The mournful cry of the loon (*Gavia immer*) greeted the men on early mornings when they awoke and ate their meal of corn and fat, then stored their gear and set off for the long journey. After several weeks on Lake Huron, Greeley and Nuttall arrived at Fort Mackinac on Mackinac Island, situated in the narrows between Lakes Huron and Michigan. Nuttall thought the island was one of the most beautiful places he had yet

Detroit, Michigan Territory, Summer 1810

seen. The island, he wrote, "rises from the watery horizon in lofty bluffs imprinting a rugged outline along the sky."

Mackinac Island was a place of rendezvous for travelers, trappers, and hunters. It was perfectly situated between the great lakes Superior, Michigan, and Huron. Besides the scattered maverick trappers and enterprising Indians, American and English fur companies used the island as their headquarters. There was much competition among trappers to meet the demand for furs and pelts, particularly of the beaver, in Europe and America. John Jacob Astor's Pacific Fur Company was preparing an expedition down the Mississippi and up the Missouri when Nuttall arrived at Mackinac. Although Benjamin Smith Barton's plan called for Nuttall to proceed to Lake Superior and journey beyond into the wilderness northwest, Nuttall immediately realized such a plan was untenable, especially because of the jealous possessiveness of British trappers who monopolized the region. Nuttall changed his itinerary in response to the generous offer of Wilson Price Hunt, the leader of the Pacific Fur Company expedition, to join with these adventurers down the Mississippi River to St. Louis, from where they intended to ascend the Missouri River the following year.

In August 1810 Thomas Nuttall embarked on a new adventure that would take him on the same path recently trodden by Zebulon Pike, Meriwether Lewis, and William Clark.

6 The Missouri River, 1811

Thomas Nuttall, in the company of French voyageurs and American trappers of the Pacific Fur Company, descended the Mississippi to St. Louis during the late summer and early autumn of 1810. The adventurers departed from Mackinac Island in their birch-bark canoes in early August. The current of the Straits of Mackinac was swift and challenging; the French voyageurs made quick work of it. The travelers crossed the northern stretch of Lake Michigan to Green Bay. They quickly navigated the one hundred miles of Green Bay to the mouth of the Fox River, which they ascended, following its "serpentine" route.

Numerous Indian tribes still lived on the shores of Green Bay and its rivers. Nuttall gained information on the customs and lifestyles of the Winnebago and Menominee tribes. The tribes, for example, relied on arrowhead (*Sagittaria latifolia*)—so-called because of the shape of its leaves—for medicine and food. Nuttall apparently witnessed the Indians harvesting wild rice (*Zizania aquatica*) by canoeing among the stands of rice and using their paddles to, first, bend the stalks over the canoe and then, second, beat the grain from the stalks. The Indian rice "is a very wholesome & pleasant food." Indian cooks sometimes mixed the rice with meat: "In the spring season after the hunt is over & they have little or no animal food they boil it & eat it with maple sugar of which they make great quantities." Indeed late winter is the best time to collect the sap of the maple by notching the tree on cold, clear nights, then boiling the sap down to a syrup or crystal. Maple sugar was an important part of the Indian as well as the early American diet. The Indians of the Fox River also venerated stone monuments, the manitou, an action that though idolatrous Nuttall thought was harmless. He felt disturbed that some Christians over the years had seen fit to destroy these totems. Do not Roman Catholics, he wondered, have similar objects of veneration? Why is one more legitimate than the other? Nuttall saw one idol of the Menominee tribe, a carved owl, being used apparently for magical purposes.

The French boatmen carried the light canoes across the portage that separated the sources of the Fox and Wisconsin Rivers—at about 43½ degrees north latitude. Over several centuries this was the preferred route from the Great Lakes to the Mississippi River for Indians and whites. The expedition passed along the western shores of Lake Winnebago to continue its journey down the Fox, but if Nuttall took an extensive detour to explore the lake, per Barton's instructions, he did not leave a record of this peregrination. Along the way Nuttall saw mallard ducks (*Anas platyrhynchos*), hiding among reeds in shallow water, and turkey buzzards (*Cathartes aura*), which the natives called "we-nung-gay." Nuttall also reported that the Indians thought that the compass plant (*Silphium laciniatum*) was "great medicine," meaning useful for a variety of ailments, particularly asthma. Once upon the Wisconsin, the expedition made a quick descent to the Mississippi River.

The confluence of the Wisconsin and Mississippi was just below the village of Prairie du Chien, which was but a small hamlet when Nuttall saw it, though still a trading center for Indians and white traders. The town had gotten its name from the Fox tribe. When Jonathan Carver, the English explorer and associate of Robert Rogers, passed this way in 1766 the town had been much bigger. From Prairie du Chien the canoes began the long descent of the Mississippi, with its frequent islands, unexpected shallows, contrary currents, islands, and natural dams caused by fallen trees still attached to their roots, blocking the flow of water and collecting branches and anything else sent by the current. There were many rapids challenging navigation in the stretch from Le Claire to Davenport, near where the Rock River merges with the Mississippi, creating what settlers called Rock Island, which was actually a peninsula. Here the Mississippi took an abrupt turn west, then south.

The Land between the Rivers

Soon the travelers were in a region of prairie dotted with oak trees where the Sac Indians lived. As they descended the river on beautiful autumn days they passed the mouths of numerous rivers, such as the Iowa, Skunk, and Des Moines, that fed the massive Mississippi. Just below Fort Madison, founded on the voyage of Zebulon Pike in 1805, they came to another set of rapids, near the Des Moines River. The Mississippi proved contrary in other ways besides the rapids, not only hidden sandbars but also sawyers—floating and bobbing logs—and planters—secure logs just below the surface. Eventually the expedition came to a beautiful, dangerous spot in the river where the Mississippi met the Illinois River to the east and the Missouri River to the west. The explorers knew that St. Louis was just a short distance away.

At the end of September they arrived at St. Louis, a growing city of over a thousand inhabitants, situated on bluffs looking over the Mississippi from the west. The Missouri and Illinois Rivers were just a few miles to the north. St. Louis was perfectly situated to dominate the trade of the Louisiana Territory and its subsequent political and territorial manifestations. Whereas many settlements came and went because of contrary changes in the route of the Mississippi, the limestone bluffs of St. Louis provided a stable trading outpost. The Frenchman Pierre Laclede had founded St. Louis in 1763. Its inhabitants were still largely descended from the French; there were also hundreds of slaves. Auguste Chouteau and his sons and Manuel Lisa were local leaders.

St. Louis was a collecting point for the adventurous and ambitious, which sometimes included scientists such as Thomas Nuttall and John Bradbury. By chance Bradbury, an English botanist who was working for Liverpool scientists trying to obtain specimens and seeds of American flora, was in St. Louis preparing to journey west. Bradbury and Nuttall met, and by further happenstance both joined the same outfit. Although Bradbury had planned a journey to the Arkansas valley, he changed his plans to accompany the Pacific Fur Company explorers on their journey up the Missouri.

Between October and March Nuttall explored the vicinity of St.

The Missouri River, 1811

Louis, waiting for the expedition to begin. He visited the Cantine (Cahokia) mounds east of St. Louis in Illinois. Nuttall described them as "100 ft. high . . . , perhaps the largest & most perfect specimens of Indian fortification, the largest of which is about 2500 ft. circumference. The area of some of them are now occupied as gardens by the monks of La Trappe"—Roman Catholic Trappist monks. The mounds "are like all others which I have ever heard of, strewed with fragments of earthen ware & human bones." He botanized as much as he could during these winter months and visited the garden of Auguste Chouteau. Bradbury, perhaps, introduced him to other notables, such as William Clark, and showed him other botanic and geologic wonders as well.

The band of guides, hunters, voyageurs, and trappers led by Wilson Hunt, Ramsay Crooks, and Donald McKenzie began the ascent of the Missouri on March 12, 1811, with John Bradbury and Thomas Nuttall in tow. Hence the Yorkshire lad found himself in 1811 exploring Louisiana in the wake of Lewis and Clark, mingling with the likes of the adventurer John Colter, the naturalist Henry Marie Brackenridge, and the fur trader Manuel Lisa. Pierre Dorion, accompanied by his Indian wife, served as a pilot, as he had for Lewis and Clark a few years before. The voyageurs were French-Canadians who sang jolly songs as they rowed the keelboat upriver. Bradbury and Nuttall went on shore to botanize on numerous occasions. The cottonwoods (*Populus deltoides*), which Nuttall would find in 1819 dwarfed on the Arkansas River, were along the Missouri massive and dominating. Bradbury recorded in his journal that some were over "seven feet in diameter, . . . with a thickness very little diminished, to the height of 80 or 90 feet, where the limbs commence." The greater horsetail (*Equisetum hyemale*), a thick cane, dominated the wet shore, which often made exploration difficult. They observed and studied buffalo, pigeons, rattlesnakes, prairie dogs, and beavers. Early in the journey they saw an old man standing on the riverbank who turned out to be Daniel Boone. "He informed me," Bradbury wrote, "that he was eighty-four years of age; that he had spent a considerable portion of his time alone in

47

The Land between the Rivers

the back woods, and had lately returned from his spring hunt, with nearly sixty beaver skins." Nearby lived John Colter, who having traveled with Lewis and Clark and having had many adventures fighting with the Blackfeet and other Indians of the upper Missouri, told his tale to all who would listen.

On March 21 the two wandering botanists became separated from the boat and were forced to contrive a passage across the Missouri. Bradbury discovered that Nuttall could not swim. Nuttall was not alone in this strange phenomenon of men who spent so much time around water not knowing how to survive in it. Most French voyageurs likewise did not swim. Hunters, however, almost always did, it being necessary to sometimes traverse rivers by means other than wooden boats. Bradbury offered to carry Nuttall on his back, but Nuttall wisely refused. So the two scientists hiked until they found, serendipitously, an abandoned raft, which they used to return to the boat.

John Bradbury provided Thomas Nuttall with a wonderful education in the ways of the Plains Indians. Bradbury was Nuttall's senior by about fifteen years and did not want the younger man to forget it! Teacher and student interviewed and interacted with a variety of Indians, including the Osage, Maha, Sioux, Poncar, and Aricara. There was much bluff and counterbluff with the Indians; one could go from apparent friend to enemy in the space of a few hours. They traded, heard speeches, and smoked the peace pipe, or calumet, the substance smoked being kinnikinnick—a mixture of dogwood, the "bark of *carnus sanguinea*," which can have a narcotic effect; and the "leaves of *rhus glabrum*," smooth sumac, a standard and useful medicine for, ironically, asthma. The shaman of the Aricara tribe smoked kinnikinnick as part of a religious ritual, blowing the smoke upward to the Great Spirit. Because Bradbury and Nuttall were interested in flora and collected samples and seeds, the Indians regarded them as "medicine men." Nuttall and Bradbury in turn respected the Indian shaman's knowledge of materia medica. The Aricara shaman carried a pouch of healing herbs, which included the "down" of the cattail (*Typha latifolia*), useful for treat-

The Missouri River, 1811

ing burns and other skin problems; wormwood (*Artemisia*), useful for intestinal problems and malaria; and *Rudbeckia* (black-eyed Susan and green-headed coneflower), used to treat burns, skin irritations, snake-bite, and the like. The scientists observed war dances and tried to converse with the Indians in bits and pieces of French, native dialect, and hand signals. Bradbury began a lexicon of the Osage dialect. Nuttall, already fascinated with Indian language, custom, and history, became completely immersed in the ethnographic study of the Indians (Aricara and Mandans) of the upper Missouri (North and South Dakota).

While Bradbury returned to St. Louis, Nuttall spent the summer of 1811 at the trading post situated at Fort Mandan. He ranged far and wide around the upper Missouri valley, approaching the Canadian border, traveling west far enough to see what he thought were the distant foothills of the "northern Andes," the Rocky Mountains. At one point the preoccupied naturalist got lost and almost died save for the friendliness of an Indian. Perhaps it was Nuttall's reputation as a shaman that saved him. Overall at the Mandan villages Thomas Nuttall received a full education not only on the customs and characters of the Plains Indians but also on their foods, agricultural practices, and medicinal herbs.

Thomas Nuttall, having done his best to follow Barton's instructions (though he had realized with experience their utter impracticality), descended the Missouri River during the autumn of 1811 in the company of Manuel Lisa and Reuben Lewis, Meriwether's brother. The journey to the Mississippi River then New Orleans was quickly accomplished, Nuttall arriving with a satchel filled with botanical specimens and seeds. Along the way he had experienced a remarkable phenomenon, the flight of thousands of whooping cranes (*Grus americana*), "a vast migration . . . from all the marshes and impassable swamps of the north and west. The whole continent," Nuttall recalled in *Ornithology*, "seemed as if giving up its quota of the species to swell the mighty host. Their flight took place in the night, down the great aerial valley of the river, whose southern course conducted them every instant towards warmer and more

The Land between the Rivers

hospitable climes. The clangor of these numerous legions, passing along, high in the air, seemed almost deafening; the confused cry of the vast army continued, with the lengthening procession, and as the vocal call continued nearly throughout the whole night, without intermission, some idea may be formed of the immensity of the numbers now assembled on their annual journey to the regions of the south." One hundred years after Nuttall wrote these words, the whooping crane would be almost extinct.

Nuttall passed New Madrid about a month before the tragic earthquake destroyed the town. John Bradbury, having spent several months in St. Louis recovering from illness, arrived at New Madrid just as the earthquake occurred, in mid-December 1811. Meanwhile Nuttall departed from New Orleans just before Christmas. War was on the horizon, and the Englishman Nuttall felt compelled to return home. There he spent three long years during the War of 1812 studying, arranging his collections, and preparing for another excursion to America. Peace brought him back to Philadelphia in 1815, eager to pursue his avocation.

7 Missouri Territory, 1815

The heat of the day, the sun of unequaled brilliance, blinding and hot, drove life underground to cool earth; or under the shade of meek cottonwoods and thirsty willows; or to shallow ponds surrounded by stands of cane; or to rivulets where muddy, tepid water trickled between the stark banks of red soil hosting countless undistinguished shrubs and grasses. A few creatures gloried in the heat. Locusts cried their shrill, repetitive song. Lizards and scorpions darted among rocks and bushes in search of prey. Ticks, fleas, and other vermin searched for available hosts. Otherwise all was still save the wind scouring the land, waves of hot air gusting from an invisible source, bending the trees with its force, flattening the grass, elsewhere churning up sand and dust. The roar of the wind was constant. The rays of the sun, almost directly above at noon, inched slowly west with the passing of day. Late afternoon brought billowing clouds building upward and outward. The rolling mass, reflecting the light of the late-day sun, rose to an astonishing height. Observation gave way to imagination as the massive anvil grew closer and to awe as it darkened the sky and halted the hot southern wind. A brief calm, and distant rumbles exploded into a sharp north wind and claps of thunder, lightning piercing all in its path. The glare of the noon sun was a distant memory during the twilight of the day. Massive thunderheads, white and gray mushrooms in the sky, filled the horizon. Dusk had an eerie, green look about it. The fascinating show of light streaking across and within the thunderheads became the immediate peril of dancing bolts of lightning, sporadically and unpredictably erupting into terrifying concussions of sound and light. Torrents of rain and hail blasted shrubs and turned the driest earth into sticky mud. Rushing runoff had its way

The Land between the Rivers

with the soil, creating deep furrows, eroding a path toward neighboring gullies and instant creeks that quickly became swollen with the dull, red water.

The deafening booms retreated; the flashes of light became less terrifying; the rain and wind lost ferocity. Calm returned again to the prairie. The utter darkness of the still, cloudy night oozed with moisture and humidity. The crickets began to chirp in earnest, their chorus growing louder by the minute. All was cleansed by the storm. Peace prevailed again. Slumber embraced the world in the hours before dawn, when the daily cycle of August on the prairie would begin again.

One would like to know what the hunter thought of the display of sound and light. Certainly he was used to the storms of the prairie, with their paradox of horror and danger yet welcome relief from the brutality of the late-summer sun. Strange, too, was the hot southerly wind that brought the heat yet at the same time dried the sweat from the brow. Streams and rivers typically provide sanctuary from the rays of the sun; so, too, do groves of trees: sweet-smelling pines, towering oaks. Such was this land, however, that it turned the typical into the unexpected. Broad rivers—the Arkansas, Canadian, Red—crisscrossed the landscape, but who could drink from their warm, salty waters? And besides the canes and cottonwoods, rarely were cool forests found, and these were to the south, where the air was warm and humid, or in the highlands of the region, requiring steady feet or a sturdy mount. The land's extremes kept the hunter on the watch, wondering about his next meal, searching for potable water, examining the western horizon—north and south—for signs of storms in summer and cold fronts in winter. One day the leather shirt and trousers felt just right for the cool breezes and gentle sunshine; the next day would bring frost and searing winds, followed by a day of humid rain.

The climate of this western territory was, in short, hard to read, even for a man who had spent years traversing its plains, crossing its rivers, ascending its hills, trapping its ponds. The hunter, whose name was Lee, came to the land between the rivers, the Arkansas and the Red, in 1810. He was one of a handful of Americans willing to compete with the Osages, Cherokees, and Comanches, as well as the French voyageur and the Spanish soldier, for beaver pelts, buffalo robes, and the right to hunt and use the land. In time this place of extremes would be called Oklahoma; to Lee in 1815 it was the southern Missouri Territory, formerly the Louisiana Territory, recently purchased in 1803 by the American government. It was open for hunting and trading as long as one minded one's business and kept on the move.

Lee could scarcely care about politics, the movement of peoples, the contests of great nations, though sometimes his own movements—his quest for survival, pelts, and wealth—were impeded by powers, governments, and people greater than himself. His was a solitary world. Contacts with other hunters, though rare, were the most welcome. The chance meeting with a squatter deep in the woods on a cold winter night could also be pleasant. Business demanded frequent visits to government trading posts, factories, or posts and taverns run by individuals. Survival often required purposeful visits to native Indian tribes—the Osages, Cherokees, Quapaws, Choctaws—for supplies and food. Yet meeting with a band of warriors coming or going was never pleasant. On their way to war such warriors considered anyone in their path a trespasser. Returning from victory, flushed with success, scalps still dripping with blood, their minds full of killing and looting, warriors hardly respected the concerns of the hunter, his pelts, horse, and gun. Worse still was the band returning with their goals of conquest and killing unmet, their thoughts intent on bloodletting, their pride demanding trophies for all to see, for which to dance and sing the songs of war and victory. Lee had his run-ins with such war parties, but so far he still had his life, paid for with pelts, horse, weapons, clothes, and pride. The memory of wandering into a post naked, hungry, cold, and sore from repeated blows filled Lee's mind and directed his actions when alone on the hunt. Silence; inconspicuousness; patience; stillness; faith; courage; wit; quickness; knowing when to give up, retreat, hide, or accept punishment without flinch-

The Land between the Rivers

ing: these were the characteristics required of he who would survive the Missouri Territory in 1815.

The hunter Lee feared the Indians—yet respected them too. Few of his activities did not depend directly or indirectly on what he had learned from Indian culture and society, especially their movements in, understanding of, dependence upon, respect for, and ability to survive in nature. The natural environment challenged as it relented, took life even as it gave life, guarded in apparent jealousy its gifts yet at the same time presented the means of survival, terrorized and comforted, forced action and gave rest. Lee received great benefits, food and shelter, from the Osages yet knew their inconsistency—hence their purposeful aggression, spirited vengeance, and spontaneous violence—as well. He relied on their trade even as he feared their thievery. He knew well Osage paths across the prairies, through the forests of the Arkansas and Red river valleys, traversing the Kiamichi Mountains. These paths had been used by Osages or other tribes for centuries, millennia. These traces were paths of survival—Lee treated them with the honor they were due.

Though the Indians could be enemies and infrequent friends, Lee respected their customs, culture, and institutions. Visiting the Osages he brought gifts, typically tobacco, as a peace offering and would never refuse the commensurate calumet, the pipe smoked to engender peace among all present. When the meal was offered, no matter the extent or quality, it was humbly accepted as a token of hospitality, as well as a humble offering of thanks to the Great Spirit. Lee had eaten his share of pumpkin marmalade and was the better for it. The product of Christian society, Lee's beliefs were rudimentary, his faith grounded in survival, his worship limited to feelings of awe upon viewing another glorious sunset, his piety firmly based on a warm bed and full stomach. The Great Spirit, the Father of Life, so said the Indians, furnished all of these blessings and many others besides. Lee was not wont to disagree. He could see the benevolence of God reflected in the rivers, prairies, forests, and mountains, just as the Indians saw the reflection of the Great Spirit in the manitou, the holy spot of singular wonder found in nature.

Hunters such as Lee perused the Indian encyclopedia of knowledge written into their practices, habits, and customs. The best means of travel in America was by canoe, particularly those made of bark from the white birch tree. The scarcity of such trees in the Missouri Territory, however, required hunters, Indian or white, to use an alternative material source. Indians along the banks of the Arkansas discovered that the cottonwood tree was soft enough to furnish a suitable dugout canoe, if something better (yet rarer), such as the gum tree, could not be found. Ash made the best paddles. Lee constructed canoes according to the needs of the moment and with whatever materials were available.

Mr. Lee often eschewed river for overland travel. Such was the ongoing demand for beaver pelts in the civilized East that Lee made trapping his primary occupation. Numerous small creeks drained and watered the forest, mountains, and plains of the Missouri Territory. Clear, cold water spilled from the Ozark and Ouachita Mountains, creating bold and rocky streams that coursed through the land seeking larger river paths, eventually the Arkansas or Red. Further west muddy creeks furrowed and drained the soil with barely a trickle during summer months—though they could change to crashing torrents in moments, depending on the rains. Inevitably streams and creeks spread their waters into small, beautiful wilderness ponds. Stands of oak, cottonwood, and willow marked the path of water, particularly on the plains. These standing groves were Lee's stopping points on the trail. Water brought game such as the white-tailed deer, an astonishingly beautiful creature preyed upon by wolves, bobcats, and hunters like Lee. Dried, "jerked" venison was often his daily fare. A fresh kill meant a haunch roasted over fire and a delightful meal.

Like most hunters Lee wore buckskin trousers and shirts, which provided serviceable protection against insects, rain, and cold— though buckskin could be unbearable on hot humid days and uncomfortable when fully wet. Lee's buckskin came into contact with enough briars and thorns that it required constant repair. So, too, did his moccasins, made of tough deer hide. One day on the trail was enough to convince the most stubborn greenhorn of the efficacy of moccasins. The hunter was constantly wading through swamps and bogs and shallow or not so shallow creeks and rivers. Moccasins dried quickly. They kept feet warm in winter. The inevitable barriers of the trail—rocks, roots, holes, mud, muck—were less daunting to this flexible leather footgear. There were no cobblers on the trail—nor were any required. Lee spent many nights by the fire repairing his moccasins with the endless supply of buckskin provided by the forests and pond groves of the wilderness.

The reason, of course, for the many ponds of the Missouri Territory was the beaver. This aquatic animal had a reputation among American hunters for sagacity. Venerable furred survivors of beaver ponds had figured out the techniques of the trapper and had passed along the wiles and ways of beaver wisdom to their young. This education, according to hunters of the Missouri Territory, extended even to the proper techniques of dam building, such as the best candidates among trees to gnaw and fell exactly across a stream, the furry engineers bringing mud and sticks to form a dam. Like its cousin the muskrat, the beaver benefited from an oily musk that kept its coat proof against dampness and cold. The beaver's skin was dark brown, its legs short and stubby, its tail flat and wide, its teeth long and sharp. Beaver ponds dotted the wilderness, the product of the beaver's unending labor to mold its environment. Beavers were clever, suspicious animals—yet jealous of their own kind as well. The hunter used this to his advantage. He would bait his iron trap with the musk oil, which attracted prey intent on driving away its rival, only to die in the process.

Buffalo hunting was made easier when the hunter knew the typical paths of Indian tribes out on their summer hunts. The numbers of buffalo present on the prairie gave off such an eerie "vapour" and rank smell that the hunter knew where the herd was even when it was hidden by an intervening bluff. Hunters learned that the best

Missouri Territory, 1815

morsels of the buffalo were the liver, the marrow of the bones, and the "tallow," or fat. The latter, mixed with corn not yet ripe, was delicious. So too was a meal of jerked buffalo meat and corn. Lee discovered as well that the Indian habit of raising dogs for food paid off when the harvest failed. The stingy forest could provide sufficient food if one knew where to look, as did the Indians. Many a hunter relied on the groundnut (*Apios americana*)—*pommes de terra*, as the French called it—which could be dug and eaten immediately to stave off hunger.

Indeed, the bounty of the plains and forest surrounded the hungry hunter. Some food was obvious, dangling from trees like the persimmon, which if tart when fresh and raw could be dried and used to make a fine bread. Other fruit trees dotted the Missouri Territory, offering apples, plums, berries, and the luscious papaw. Nut trees were plentiful as well, ranging from the Ozark chestnut, the pecan, and the black walnut, to the varied hickories of the forest. Some nuts were good eaten raw; others required boiling to release oil that could be used for cooking. Native tribes used the *pawcohiccora* oil from the hickory tree in cakes. For a quick energy source the hunter sought bayberry or basswood trees, which hosted honeybees and hence were known as "bee trees." Hunger might force a quick ascent of one of these trees, though the satisfaction of the hunter came at a painful price. The hunter with more time on his hands would first fell the tree with his axe, then collect the honeycomb and honey. The bark of the basswood was useful in making primitive ropes.

The hundreds of wildflowers of the forests and plains decorated the Missouri Territory during all seasons of the year. Such beauty also provided food and medicine. Indians, and hence hunters, ate the plant of the showy white trillium, which grew in the dense forest, as well as the root of the delicate spring beauty, found where oaks and hickories were plentiful. The "Indian turnip," or jack-in-the-pulpit, when cooked provided nutritious food. The ubiquitous blue violet made a good salad, especially if topped off with one of the many varieties of edible mushrooms that grew in forests and

The Land between the Rivers

meadows. A few were deadly poisonous, as the Indians and not a few hunters discovered by experience.

The Indian materia medica, often scoffed at by white society, was nevertheless useful and a worthwhile study for a man alone in the wilderness. The Indian medicine man might pass along to the hunter suffering from the ague his knowledge that the bark and roots of the dogwood, which flowers beautifully in the spring, bring relief from the fever and chills. The medicine pouch of many a native physician contained the powder of the cattail, used for burns, as well as mint leaves, which hunters called hyssop, a universal tonic. The oil of the bitternut hickory, as well as the sticky bark of the slippery elm, helped relieve the aches and pains of rheumatism. The latter was a remedy for the cough, as was the bark of the black cherry tree. The bark of the wahoo, or eastern burning bush, according to Indian naturalists, was useful in purging the body of wastes. Natives had discovered as well the many roots, barks, and berries of a variety of forest and meadow trees, shrubs, and flowers that could produce medicinal teas for the sick. The lavender bergamot made a tea for the asthmatic sufferer. Tea made of the bark and twigs of the sassafras tree had long been thought of as a cure-all; even in Lee's time sassafras tea helped the hunter who was feeling under the weather. The bright red berries of shining sumac could make a tolerable drink; the young sprouts of the smooth sumac when chewed gave precious moisture to the hunter who chanced to find himself in a waterless environment.

Hunters typically sought their living in the Missouri Territory under a license from the territorial government. Perhaps Lee did, too. Perhaps Lee also reflected on the irony of a government licensing the most fundamental of human activities. The hunter's life was dangerous yet free. It made little difference whether or not a hunter was licensed when trying to navigate rapids in a dugout canoe or standing wet and shivering in the hollow of an old cottonwood tree with torrents of rain and shafts of lightning threatening at arm's length. Governing a wilderness is hard work. Nature hardly accommodates human rules and regulations. The good intentions of legislators vanish in the face of the natural law of the forest and prairie. The hunter knows intuitively the self-evident truths of equality and liberty. Nature dwarfs man. The river rushing forth overwhelms all in its path. The unexpected bends and spontaneous hazards reveal a logic of descending water that no canoe can predict. Reason breaks down when the current brings one to an impossible situation from which there is no escape. The hunter cannot know why the frigid north wind of today follows on the heels of the sticky humidity of yesterday. One cannot prepare for the apparent whims of climate and geography, the unexpected confrontations with predators—human or otherwise. Hunters like Lee had to be prepared for every contingency, knowing full well that such preparedness was never enough. The Missouri Territory required from the hunter not struggle and competition but acceptance. How can one control what cannot be controlled? Resistance was futile. One had to give oneself to fate and obediently follow where it led. The hunter was nothing compared to everything that surrounded him. He who gave in survived. The valiant dead who fought for life ironically lost it.

By habit, inclination, and utter necessity Lee and his kind mirrored the wilderness. His skin clothing from head to toe, long hair, and general unkemptness made him appear as savage as the most savage predator of the forest. He might not devour his prey alive, as the black bear of the Ouachita forest was said to do; nor did he swallow whole his conquest, as the diamondback rattler and great horned owl did their prey; nor still was he wont to isolate the young, weak, or sickly from the herd to pounce upon and tear limb from limb, as did the bobcat. The savagery of Lee was subtle. Lee's pleasures were the simple pleasures of the wild—a good fire, a full stomach, dry clothes. His bed more often than not was a pile of leaves and twigs made dry and warm by his buffalo robe. He produced fire by striking flint to send a spark into dry grass. He drank water, ate berries, munched on nuts, and looked forward to fresh roasted venison for dinner. His movements were as silent as the forest. His five senses guided his path. His map was the moss on trees, the pattern of the stars, the direction of the wind, the angle of the sun's rays.

The Land between the Rivers

Traces of the paths of deer, bear, elk, or Indian hunters led him as well. So, too, did he find direction in the paths and flow of streams, all heading in the same eventual direction—the lowest point, the valley, the river, the sea.

The hunter Lee rarely spoke, yet he could be noisy at times, particularly when mimicking a squirrel's chattering, a buffalo's lowing, a jay's squawking, a turkey's gobbling, or the rustling of a bear searching for supper. The forest is rarely silent—the wind blows, trees rock, birds sing, water drips, leaves rustle. But these are the sounds of savagery, of the wild and untamed, of nature. Lee knew these sounds, counted on them to make sense of his day, to keep track of time and seasons, to know that all is well. At rest near a pond, or on the banks of a stream, or in a gentle meadow, he listened to the savage music of a dozen birds singing at once out of harmony, the insanity of the mockingbird countering the pleasant song of the cardinal. Yet it all made sense to Lee. It all seemed right and orderly. All was good.

Lee's life as a hunter in the Missouri Territory was one continual journey. His travels through time were not hurried but were as constant as the rising and setting of the sun, the changing phases of the moon, the transition from winter to spring, summer to fall. The light of day, changes in vegetation and temperature, his waning strength and growing hunger, his movement from place to place kept time. Lee's was a life of movement simply because to halt for more than just a moment was to exhaust the land of its food, to give up stalking the game, to allow the escape of prey, to let the cold numb and the heat grow feverish. Constant movement was required of Lee and all other creatures he came upon in his journeys. River paths and traces through woods grew in familiarity with the passing years. The hunter must continually retrace his steps of before, of yesterday, of last year, to find the buffalo returned, the birds nesting again, beavers damming another stream or repairing their work from a previous year.

Lee's restless movement throughout the Missouri Territory led him to ascend and descend its rivers and streams, to journey

Missouri Territory, 1815

through its mountains and forests, to cross its prairies. Rivers and streams north of the Arkansas usually boast clear, fresh water moving from highlands to the valley. The Illinois River is such a stream. This beautiful river took the hunter Lee from the wooded mountains of northwest Arkansas, winding through oak and hickory forests, to the Arkansas River. The Illinois valley was excellent land for trapping; it paid the hunter who took the time to make a sturdy dugout canoe big enough to hold himself, his supplies, and the many pelts he expected to acquire along the way. Water descends quickly on the Illinois and hence rarely freezes; hunters could use this river year-round, but particularly in the late fall and winter, when animal coats are full, more valuable. The river's channel is seldom more than a hundred feet wide or very deep. The bottom contains sand, rocks, and pebbles. The river's depth is variable. Low water required the hunter to form an ad hoc portage and drag his canoe to deeper water. Otherwise the current is pleasant—a quick and easy descent with little to threaten or cause concern. Yet any river is unpredictable. One does well not to take the clear deep channel, the lack of significant white water, for granted. Spring rains swell the Illinois, adding to its danger; the river leaves behind signs of its path of destruction. The water of the freshet jumps its banks, destroys trees, and contorts rocks and branches into astonishing mosaics. Especially at times during or after a freshet even peaceful and passive rivers like the Illinois have their treacherous moments. Around any bend can be obstructions of trees massed together, forming an impenetrable shield; or sunken logs rooted to the riverbed; or rapid, deep water spilling into a narrow, rocky channel, forming whirlpools that toss and turn the canoe at will. Such dangers are rare on the Illinois, yet a few exist now and did at Lee's time. It took an experienced boatman to navigate even a simple stream such as the Illinois from its source to its mouth.

The Illinois's equal in beauty, and in the possibilities of the hunt, particularly of the black bear, beaver, and white-tailed deer, and hence equally attractive to the hunter Lee, was the Ouachita (Washita) River, which flowed south from its hilly source in central

The Land between the Rivers

Arkansas to the Black River, then on to the Red. The Ouachita brought from the mountains clear, good-tasting water attractive to both man and beast. Deer came as well to the salt deposits along the way. There were few obstructions on the Ouachita, so it was easy to ascend in a good dugout canoe. Lee carried the materials necessary to construct a dugout should the need arise. His equipment included an ash-handled axe for felling a good cottonwood and an adze for cutting into the soft wood.

The Ouachita, Poteau, and Kiamichi valleys were the land of the black bear. The hunter sought the bear in early winter when his fur was thick and he was still fat from his autumn repast. Bear fat, which if boiled produces oil, was in high demand in the settlements for its varied medicinal, epicurean, and household uses. Black bears are huge, lumbering creatures that make easy targets for marksmen. Yet they can be overwhelming in close quarters, unstoppable if wounded or protecting their young, and ferocious when gorging on a fresh kill. This ferocity was heightened by the legend that the bear ate its prey piecemeal, when still alive, ignoring the agonizing screams of the victim. Fortunately the bear was omnivorous, willing to eat anything besides man, including insects of all sorts, berries, nuts, fish, and rodents.

Mountain rivers in this region, such as the Kiamichi, are clear and bold, unlike the slow and winding Red, Canadian, Cimarron, and Arkansas to the southwest and northwest. Hunters who were also experienced anglers caught tasty meals of bass, crappie, bluegill, and catfish in these crystal waters. The water was potable, unlike that of other rivers to the west, fed by numerous streams, many of which attracted beaver to their waters to dam, build lodges, and live. It was Lee's job to find out where. Cane and brambles were ubiquitous, as were the wretched vermin of the forest and hills. Ticks could cover the hunter's leggings in a moment; mosquitoes hunted in huge swarms. Snakes were plentiful, especially in late spring, summer, and early fall. Lee preferred to hunt in winter months when the most venomous snakes—the rattler, cotton-mouth, and copperhead—were hibernating; when insects were nowhere to be seen; and when his vision was unobstructed by the lush green undergrowth of the spring and summer forest.

Many times Lee followed the Canadian and Cimarron far to the west, almost to their sources. The Arkansas, home to the Osages, he preferred to avoid. The Osages had a somewhat deserved reputation among hunters for duplicity and aggression. It was best to avoid an encounter—not that confrontations with other Indians of the region were pleasant. The Cherokees and Choctaws were becoming more frequent visitors to the land between the rivers, as game gave out in their former hunting grounds east of the Mississippi and due west, between the lower Arkansas and Red Rivers. Indian hunters considered white hunters as rivals to be intimidated and forced out. The Osages, for example, were very particular as to who they allowed to trap and trade in the Arkansas valley. For years they allowed Auguste Chouteau this privilege; more recently Joseph Bogy had obtained the honor. Both men were French. The French voyageurs and missionaries from Canada had for centuries treated the Native Americans with respect. English and Americans, more haughty, never earned the esteem granted to the French. American hunters such as Lee entered unfriendly territory when trapping and hunting in this wilderness land between the Red and Arkansas Rivers.

If the hunter Lee ever reached the sources of the Canadian and Cimarron Rivers, which is doubtful, he could have seen in the distance the great Shining Mountains, called variously the Sand, Mexican, Western, and eventually the Rocky Mountains. Many hunters and explorers claimed to have voyaged to the source of such rivers, including the Arkansas and the Red. The late-night campfire, the orange flames leaping amid darkness, creates contrasts and mystery, a wonderful setting for spinning yarns. Trained surveyors armed with maps, compasses, sextants, theodolites, barometers, thermometers, and the like would time and again find themselves bewildered by the maze of trails amid mountains, deserts, and prairies, whence derive the sources of the Red, Canadian, Cimarron, Arkansas, and Platte Rivers. Lee nosed and felt his way west. He

The Land between the Rivers

might not have known the elevation and latitude, the names of this creek and that hill, but he knew generally where he was. He knew the lay of the land and where a spring or pond should be. The diversity of nature, the apparent randomness of climate and topography, the confusing habits of western rivers, invited not reason and science, the vain attempt to conquer through knowledge. Why confront the land as an outsider? Lee was an insider. His approach to the river was to be at one with it, allowing the current to guide his simple canoe, using the paddle sparingly and just to conform to the river's will. Lee rarely forged new trails. What was the sense in that, when nature forges its own? The deer knows the best path through the forest. Find its trace and follow it. Eat and drink when the opportunity presents itself. What animal of the forest or plain eats three square meals a day? When the kill is made, eat hearty, fill the stomach, for the next meal might be days away. Nature rarely indulges in constancy of temperature and humidity. Gird yourself, then, for the coldest of days followed shortly by the hottest of days. Nature is unpredictable. Life is unpredictable. Why struggle against it? The rich will soon be poor. The wise in time face ignorance. The only certainty is death. All else appears random. Even an uneducated man such as Lee learned what few philosophers ever learn— that one must accept who, where, when, and what one is. To live appropriately is to live simply, not to resist life but to accept its mystery, contrariness, and perplexity. Lee possessed none of the civilized virtues of wealth, learning, power, or fame. Yet he had God's greatest gift: life. At least for the moment . . .

8 Cumberland Gap, September 1816

Thomas Jefferson, writing of the Appalachian Mountains in his *Notes on the State of Virginia* in the early 1780s, thought that the highest peaks of North America were his own Blue Ridge Mountains of western Virginia. Besides the tendency of Jefferson to assume that anything American, particularly Virginian, was grander and greater than could be met with elsewhere, his erroneous impression of the Blue Ridge revealed that eighteenth-century Virginians thought that the Appalachian chain was an almost impassable barrier to the West. This made Thomas Walker's feat in 1750 of forging a path through the Cumberland Gap all the more remarkable. Walker was a physician who left behind a written record of his journey. Most of those, however, who crossed the Appalachians through the Cumberland Gap were Indians and white hunters. Daniel Boone, for example, the most famous hunter and trailblazer of the late eighteenth and early nineteenth centuries, knew of the many Indian traces through the remote mountains of North Carolina and Virginia. One of the most famous was the Great Warrior's Path, which Boone and half a dozen men took in the spring of 1769 west through the Blue Ridge, across the Holston, Clinch, and Powell Rivers, to the Cumberland Gap.

By the time Thomas Nuttall journeyed through the Cumberland Gap it was no longer unknown, yet it was still very much a wilderness. Nuttall's journey through the gap was part of two journeys he undertook from 1815 to 1817 through the American South. The first, from 1815 to 1816, took him to Georgia, South Carolina, and North Carolina, traveling up rivers, collecting specimens, and befriending southern botanists. His second journey began in Pennsylvania, which he traversed to the Allegheny River, down which he

The Land between the Rivers

journeyed to Pittsburgh. From Pittsburgh Nuttall descended the Ohio River to Cincinnati, from which he journeyed south along a land trail to Lexington, which was at this time a leading trade center of the trans-Appalachian region. In September Nuttall journeyed south, exploring the Kentucky and Green river valleys, then following parts of the old wilderness road to the Cumberland River. From the Cumberland he wandered east through the gap, a beautiful notch between mountains that when Nuttall traversed them were just starting to reveal the autumn colors of the dominant deciduous forest. Emerging from the gap, Nuttall took the Great Warrior's Path across the Powell, Clinch, and Holston Rivers, then turned south toward the Great Smoky Mountains and the French Broad. He skirted Mount Mitchell to its east but ascended several peaks, notably Table Rock Mountain and Roan Mountain, one of the highest east of the Mississippi.

The mountains of America clearly fascinated Thomas Nuttall. Yet he is not well-known in the history of mountain exploration. In his early journeys he was unable to penetrate the Rockies, only seeing them (perhaps) from a distance. His brave attempt to ascend the Arkansas to its mountainous source in southern Colorado failed. A few years later he did succeed in ascending one of the highest peaks in America, Mount Washington in New Hampshire, although by then the route to the top was well-known and other botanists had already reached the mountain's summit. Even so, Thomas Nuttall arguably visited, recorded observations on, ascended, and traversed more mountain ranges than any other American scientist up to his time. These included the White Mountains, Allegheny Mountains, Ouachita (Kiamichi) Mountains, Great Smoky Mountains, Blue Ridge Mountains, and then, in the 1830s, the numerous mountain ranges that make up the Rocky Mountains in Wyoming and Idaho.

Nuttall typically described mountains in romantic terms of sublimity, awe, and grandeur. His drawings of the mountains of Arkansas highlight the pastoral—humans living in harmony with nature. One understands human experience in terms of the immensity of the land next to the insignificance of man. Nuttall's

Cumberland Gap, September 1816

character and self-image derived from his perception of natural and human history. An Anglican, he adopted the position of the great philosophers of the Church of England, such as Richard Hooker, that God's creation is presented to humans to care for as stewards and to learn from as the ultimate expression of that creation, formed in God's own image. One senses in Nuttall's writings that he felt overwhelmed by the responsibility entrusted to him, to all humans, to know this great gift of nature. It was not a choice for him but rather a responsibility, a duty incumbent upon Thomas Nuttall as an Englishman, an Anglican, and a scientist. He did not take the responsibility lightly but embraced it fully. Desire to explore, to discover, to know burned within him.

But when one self-perceived task of exploration, discovery, and knowledge was fulfilled, there was no time for rest, because other tasks beckoned. Nuttall felt most alive when on the trail, when alone or with an empathetic companion in science, glorying in the beauty of nature, recording the data inherent in the soil and plants, then moving on to something else altogether new. When he was at rest, halted unavoidably, enclosed within walls, forced to work, teach, converse, socialize—act, in short, like a civilized nineteenth-century English-American, he felt out of place, despondent, even depressed. He once referred to this state of mind brought on by inaction as ennui, a "weariness of mind." The best medicine for ennui, he believed, was "gentle traveling on foot," "gentle" referring not to soft and easy traveling but to slow and methodical traveling no matter the terrain or the danger. So he practiced the art of healing the restless, perfectionist, despondent mind—over and over, decade after decade, mile after mile. The pedestrian Thomas Nuttall was one of the great travelers and explorers of his own time—but to him the journey in search of knowledge was never enough, never completed, never quite fulfilled.

9 The Ohio River, 1818

Thomas Nuttall was a restless man. Not content to remain in England after the War of 1812, he returned to his adopted home of America. During the three years from 1815 to 1818 Nuttall made small journeys, collected information, and attended meetings of the American Philosophical Society. He earned respect from eastern scientific circles with the publication in 1818 of his *Genera of North American Plants and a Catalogue of the Species to 1817,* based in particular on his extensive journeys throughout America from 1808 to 1812. The *Genera* showed Nuttall to be a naturalist of tremendous creative energy, a Linnaean devoted to the quest for knowledge of the apparently countless unknown species of plants and animals of the American wilderness.

Nuttall was one of those who is never satisfied with accomplishments, always on the lookout for more. The American West, and the untold numbers of plant and animal species yet to be discovered, studied, cataloged, and given appropriate scientific recognition, beckoned. Nuttall decided to journey west a second time, but further south than before, to penetrate the lands of the Southwest, the extreme borders of the Louisiana Territory, in search of new flora and varied adventures. The journey would take nineteen months, during which time Nuttall would approach the brink of his personal unknown even as he came to be fully acquainted with the terra incognita of the American southwest.

Thomas Nuttall lived in a restless age. America in 1818 was a place of opportunity for the stalwart and adventurous soul, the risktaker and entrepreneur. Nuttall's was a time when the preceding three centuries of European exploration and scientific discovery had opened doors to the vastness of creation, the diversity of nature,

The Ohio River, 1818

and the amazing variety of human communities spread throughout the world in diverse, mysterious lands. The great accomplishment of Western civilization over the preceding few centuries had been to open to the human mind hitherto unknown peoples and places; to reduce the level of human ignorance; to reorient the mind to reason rather than idle superstition; to show the manifold possibilities available for humans to discover, to know, to comprehend. Thomas Nuttall and his contemporaries, if they still had much to explore and discover, if ignorance was still their curse rather than knowledge their blessing, nevertheless enjoyed the feelings of youthful exuberance—of the door opening, of the singular purity of innocence that precedes the complexities and ambiguities of experience—that few others have since known.

In 1818, when Nuttall began his journey west, there existed the image and the reality of the Arkansas River and its source. The harsh truth about western lands—the alternating drought and drenching rain, hot winds of the south and cold storms of the north, beautiful evening zephyrs and dazzling sunsets contrasted with massive storms and killer tornadoes, flash floods and dried-up creeks, water fit only for beasts and pleasant beaver ponds, the graceful red-tailed hawk floating on air filled with swarms of mosquitoes, a mockingbird squawking at an intruder who has discovered the nest of defenseless chicks—settled uneasily on the human mind. Such contradictions of terror and beauty elicit a selective response. Americans in 1818 were hungry for land in the former Louisiana Territory, which had been subdivided into smaller territories to accommodate settlement. In the years since the War for Independence American writers and thinkers had christened the land and its peoples with romantic qualities. Hector St. John de Crèvecoeur proclaimed Americans to be "new men" created by a continent, imbued with an inherent love of freedom and equality. Nationalists such as James Madison envisioned a republic encompassing the entire breadth of the continent. Clergymen such as Jeremy Belknap saw the mark of God written on every mountain peak, crystal lake, sublime canyon, spectacular waterfall. Boosters such as John Filson

The Land between the Rivers

created images of the wilderness sage, the frontiersman mirroring the wisdom and goodness of the virgin inland forests. Artists envisioned America endowed with humanlike qualities of beauty, strength, and constancy. Indeed, for centuries Europeans had pictured America as a place of "noble savages," of rebirth and re-creation, of "a city upon a hill," the home of God's new chosen people. But then Europeans had for millennia looked west across the Atlantic, their minds discovering paradise, the Isles of the Blessed, the pastoral Arcadia—purity, wonder, goodness, truth.

To go west in 1818 was to return to the past. Across the Mississippi River was a region that had, for centuries, been designated terra incognita. The Mississippi was a time zone. On its eastern side was civilization. There one could always be sure of the time, the date, the place, the rules, the prevailing institutions, law and order, surplus goods and surplus people. The East heralded the future. Boston and Philadelphia were centers of civilization, where the American of 1818 could find the sophistication of technology, government, culture, and society. The Industrial Revolution loomed on the horizon like a tall ship heading into port. The East symbolized progress, youth, movement forward.

Across the Mississippi was a land before civilization, before history, a place in the distant past of humanity. In such a place the only boundaries are rivers and hills; the only time is the passing of night into day, winter into spring. Dates and chronology have no significance when the sun beats down on the desert and water is scarce. The latest fashions from New York or Paris are meaningless compared to the billowing thunderheads on the horizon. And what does it matter who is president or what war is occurring when the only thing between you and death is the swiftness of your horse, your wits, or the luck of the draw?

Philadelphia in 1818 could not have been more different from the trans-Appalachian and trans-Mississippi frontier of the United States. It was a city of well over a hundred thousand people, a mixture of ethnicities, languages, and customs; of decorum and rudeness, wealth and poverty. When Alexis de Tocqueville visited the city

The Ohio River, 1818

in the 1820s he was astonished by the moral filth of the Philadelphians. And yet the finer sort of people still sought to stay ahead of other American cities in terms of fashion and trends. The beauty of the rich young women of Philadelphia was talked of even across the Atlantic. Philadelphians still basked in the poetry of Philip Freneau, in the status of Independence Hall, in the memory of Franklin. Yet Thomas Nuttall could not resist the pull of the West.

The West was Thomas Nuttall's goal when he departed Philadelphia on a brisk fall day in 1818. Ahead of him lay hundreds of miles of travel through lands known and unknown. Ahead lay a journey of a year and a half west across the Appalachian Mountains, down wild rivers, beyond the frontier to the American wilderness. Ahead of Thomas Nuttall lay a journey into the past. America west of the Mississippi in 1818 was the home of French traders, unpredictable Indians, and the occasional hardy trapper and homesteader. The American West might accommodate the whims of the eastern visionary and romantic, yet it was hardly accommodative to the aims of a scientist. But what better laboratory is there than the unknown? There, Thomas Nuttall convinced himself, was virgin territory for the scientist.

The thirty-two-year-old man who set out from Philadelphia on October 2 for Lancaster, Pennsylvania, was impatient to be "pleasingly amused by the incidents of traveling." Thomas Nuttall fit well the caricature of the scientist: obsessed with details, absentminded, zealous for esoteric discoveries. He hardly seemed the type to endure the fatigue of the trail, to camp out under the canopy of stars, to guide flatboats through rapids, to mix with pirates and other riffraff, to journey to Indian Territory and sit with the native inhabitants in their tepees and wigwams smoking the pipe of peace. What Thomas Nuttall proposed to do over the course of the ensuing seventeen months filled him with trepidation. Nuttall kept a journal to record observations and scientific discoveries—and to release by means of pen and paper the anxiety and wonder of his journey into the wilderness.

From the perspective of the stagecoach Nuttall found that the val-

The Land between the Rivers

ley from Philadelphia to Lancaster "exceeds every other . . . in fertility and rural picture." The inhabitants were the descendants of German immigrants—they still spoke the language of the old country. Yearning to involve himself actively in travel, Nuttall eschewed the coach at Lancaster; as a "pedestrian" he would enjoy the "beautiful succession of hill and dale" of Pennsylvania. Besides, foot travel affords "better opportunity for making observation." On the road from Lancaster to Harrisburg Nuttall passed through "a fertile tract" called by the Shawnee Indians "Pleasant Fields." He crossed the Susquehanna River on a bridge divided in the middle by a large island. The geology of the "sylvan hills" fascinated the scientist. On a good road Nuttall made the thirty-one miles to Shippensburg in one day of energetic traveling, the forested slopes of the Allegheny Mountains before him. The beautiful, odorous berries of fragrant sumac (*Rhus aromaticum*) provided a pleasant distraction. West of Loudon the turnpike ascended a large hill from which Nuttall observed "a wide and sterile forest extending across the glen, and only at small and distinct intervals, obscurely broken by scattered farms." The inhabitants lived impoverished lives in rough cabins of varying construction. The road through these "deep and narrow valleys" and "steep hills" was "as bad as can well be imagined"—"a mere Indian trace." The town of Bedford helped to break the monotony of travel. Nuttall thought it "very pleasant and romantically situated" in the Juniata valley; covered bridges spanned the plentiful streams. A toll bridge crossed the Juniata, which "does not exempt the pedestrian traveller." The river cut a picturesque gorge dotted with "pine and spruce" trees.

At Bedford, on the route to Pittsburgh, Nuttall picked up the Forbes Road, which encouraged "convenience and facility to the inland commerce of the state." Laborers were busy providing much-needed maintenance. Hills dominated by stands of oak (*Quercus*) and pine (*Pinus*) surrounded the solitary traveler. Even so fine a naturalist as Thomas Nuttall could not engender constant enthusiasm for the "monotonous" and "gloomy" forest. Numerous Conestoga wagons loaded with goods and pulled by teams of horses provided some diversion. Also providing diversion for the botanist were the rhododendron (*Rhododendron maximum*), the cucumber magnolia (*Magnolia acuminata*), and the serviceberry (*Amelanchier canadensis*), the latter two good medicines particularly for the ague and diarrhea. Nuttall gathered seeds along the way. The chestnut oak (*Q. prinus*), its broad leaves turning to autumn colors, dominated the hills on this leg of the journey to Pittsburgh. Notwithstanding the turnpike, the few towns in the valleys were dirty with indigent inhabitants and too many "dram shops, improperly called taverns." The profusion of rhododendron on Laurel Mountain provided a wonderful contrast to the depressing prospects of town life. Indians had long used rhododendron and mountain laurel (*Kalmia latifolia*) to treat muscular and joint ailments, in spite of the plants' toxicity. Nuttall, nearing Pittsburgh on October 14, found travel "laborious," the road difficult, the towns ugly; the aroma of burning coal (used to heat cabins) competed with the cool air tinged with the smells of autumn. At length he reached Pittsburgh.

Nuttall arrived at Pittsburgh for his third visit. On the first, eight years before in 1810, he had enthusiastically approached the city with romance and expectation. "On descending the last of the high hills and approaching *Pittsburg*," he recalled, "a pleasing prospect discloses itself and forms a fine contrast with the past; the eye wearied with the barren prospect of rocky mountains, and black forests, views with satisfaction and delight the verdant meadows and smoking cottages where ere long the fierce savages held their nocturnal dances, and the ancient forests planted by nature raised their towering heads unto the skies." His experience of Pittsburgh in 1810, however, had taught the young botanist not to be so sanguine: "The streets are narrow and not very regular. The houses seem merely built for convenience, displaying little or no taste. . . . The principal fuel here is coal, which beside blackening and disfiguring the houses envelopes the city in a cloud of smoke."

Upon arriving again at Pittsburgh in 1818 Nuttall tried to be generous in his sentiments. On the one hand Pittsburgh, "the Thermopylae of the west, into which so many thousands are flocking

The Land between the Rivers

from every Christian country in the world," impressed Nuttall. The city was a focal point of trade: "The shores of the Monongahela were lined with nearly 100 boats of all descriptions, steam-boats, barges, keels, and arks or flats, all impatiently and anxiously waiting the rise of the Ohio, which was now too low to descend above the town of Wheeling." On the other hand Nuttall was too much the naturalist to forgive a city of which the predominate characteristics were "filth and smoke and bustle."

The Monongahela and Allegheny Rivers meet at Pittsburgh to form the Ohio, which upon rising became the route for Thomas Nuttall, as for so many thousands, to the West. Requiring a means of transportation, Nuttall purchased a "skiff" and hired a young adventurer who purported to know the art of navigation. They set forth on the descending Ohio on October 21, 1818. Their first day on the river ended in a thunderstorm, which required their retreat to "the first cabin which we came to, built of logs, containing a large family of both sexes, all housed in one room, and that not proof against the pouring rain." The host was "hospitable" and attentive to their needs and "would scarcely permit any of his family to receive from us the moderate compensation which we offered."

Dawn brought sunshine after the rainy night. Nuttall and his companion took to the river, the water level of which continued low, with frequent rapids. A head wind slowed their progress, at length forcing them to halt for the night at a cabin adjoining a tavern wherein was a scene of drunken revelry. Nuttall, a teetotaler wanting only repose, was disgusted and offended when the revelers retired to the cabin for dancing and drinking: "The whiskey bottle was brought out to keep up the excitement, and, without the inconvenience and delay of using glasses, was passed pretty briskly from mouth to mouth, exempting neither age nor sex. Some of the young *ladies* also indulged in smoking as well as drinking of drams."

After this brief "interrupted repose" the travelers set out again. The morning was cold; snow covered the ground. Descending the river required considerable effort against the prevailing head wind. On several occasions the cold forced the two to retire to the warmth and uncertain hospitality of log cabins along the shore. There, coal fires warmed them. One host, "an Amazon, modest, cool, and intrepid," entertained them with stories of her misadventures—of descending the Ohio to Cincinnati; losing her husband to fever; ascending with others the Ohio in search of a better home; being "struck by a hurricane" on the river; finding that she alone among the passengers, which included many males, had the courage and audacity to navigate the boat in the storm.

On October 24 they crossed into the state of Ohio. The banks of the river were "exceedingly romantic, presenting lofty hills and perpendicular cliffs." Trees, clothed "in their autumnal foliage," overwhelmed the hills and enclosed the river, the course of which appeared as a narrow path through an arbored gauntlet. Dawn on the river was a time of sublime beauty—preceding the winds of full daylight a "blueish" fog hovered about the valley like a massive roll of the densest cotton fiber.

On October 26 they came to Wheeling, a town built on the Virginia side of the Ohio River. Wheeling consisted "of a tolerably compact street of brick homes, with the usual accompaniment of stores, taverns, and mechanics." The town, almost fifty years old, had grown in part because of its situation on the National Road, the interior turnpike laid out and maintained by the federal government to insure the growth of inland commerce and communication. Scattered boulders of sienite in the neighborhood of Wheeling suggested to Nuttall's geological mind that they were "adventitious" and had immigrated—how he was not sure—from "the mountains of Canada." The next day, passing the mouth of Little Grave Creek, Nuttall halted to explore the famous Indian mound, long the focus of interest among American scientists. Nuttall, who had never seen the Egyptian pyramids, nevertheless thought the seventy-nine-foot-high Grave Creek Mound was comparable.

The urge to explore and botanize impelled Nuttall to walk along the shore while his companion—and another newcomer Nuttall had hired—took charge of the skiff. Nuttall delighted in the presence of the lofty elm (*Ulmus americana*), its elliptical leaves still a bright yellow, as well as the smooth-barked beech (*Fagus grandifolia*), its nuts perhaps tempting the hungry explorer.

The wind changed on October 30, a cold southwesterly blowing upstream, which forced the three to labor just to propel the skiff downstream. The last day of October Nuttall discovered an aster (*Aster*) still in bloom. November brought with it a heavy thunderstorm, which required lodging with one of the many desperately poor farmers of the region, who could offer the travelers only "mush and milk," with perhaps a little maple sugar to sweeten it.

On November 2, delayed again by the wind, camped on the riverbank, two dogs "belonging" to one of Nuttall's skiff mates cornered a "buck" and drove him to the river, where he was easy prey for the hunter; they feasted on venison that night. This first week of November 1818 they made slow progress due to the strong wind from the southwest blowing freely up the Ohio valley, which descended south-southwest. Constant delays allowed Nuttall to explore the environs. He found the papaw (*Asimina triloba*) in abundance, its oblong, bananalike fruit tasty; the mistletoe (*Phoradendron serotinum*), a beautiful shrub to behold, its leaves forming a medicinal tea but its fruit poisonous; and the celandine poppy (*Stylophorum diphyllum*), not in bloom, surrounded by a forest of beech (*Fagus*) and basswood (*Tilia*).

On November 7 they came to the border of Kentucky, where the river turned abruptly north. Coming to rest late in the evening on the Kentucky side, they happened upon a harvest celebration, which included "corn-husking" and other activities of "merriment and riot." Nuttall concluded that the inhabitants of the Ohio valley were rather too fond of such amusements, resulting in overwhelming poverty and a poor diet, which doubtless encouraged continued sloth. Cocklebur (*Xanthium strumarium*) plagued cattle and sheep in this region of the Ohio; for humans, however, the plant had medicinal qualities, such as treating symptoms of malaria. Here, the skiff mate with the dogs went his own way.

By the time they reached Portsmouth, a town built on the site of

The Ohio River, 1818

an ancient Shawnee village at the confluence of the Scioto and Ohio, the wind changed to a northerly, which pushed the skiff along briskly in its southwesterly descent. For two days they made good time. Then on November 11 the wind changed and a cold rain blew in, "which in our open skiff proved very unpleasant, and, to augment our uncomfortable situation, we encamped at a late hour on a very disagreeable muddy shore, where it was not possible to kindle a fire." Hereabouts, Nuttall noted, the inhabitants were industrious New Englanders who nevertheless lived in poverty in well-built framed houses totally lacking in furnishings; the houses appeared to Nuttall to be little more than barns.

Nearing Cincinnati Nuttall's companion became "insolent," that is, impatient with the fastidiousness, inflexibility, and scientific obsession of his employer. Nuttall convinced himself that the man "was a refugee from justice or deserved infamy, and in all probability I narrowly escaped being robbed." Fortunately, at Cincinnati Nuttall found a man more to his liking, the scientist and physician Daniel Drake. Nuttall and Drake were old friends, the latter having been a student of Benjamin Barton at the University of Pennsylvania.

Cincinnati when Nuttall saw it in 1818 was thirty years old. Industrious New Englanders had settled the town at the junction of the Ohio and several small rivers, most notably the Licking River, flowing north from the rich hinterland of Kentucky. To the river-goer Cincinnati presented the pleasing prospect of an amphitheater, rising in successive strata from the river. Cincinnati singularly impressed Nuttall, who thought the city was "by far the most agreeable and flourishing of all the western towns." Before departing Cincinnati Nuttall met and talked with Colonel Hugh Glenn, recently returned from the Arkansas valley, Nuttall's ultimate destination. Colonel Glenn was a merchant and adventurer hoping to establish a trading post at the Three Forks. He furnished a letter of introduction for Nuttall to present to Joseph Bogy, the French trader, to solicit his assistance.

Armed with encouragement and information, Nuttall tarried no further, setting forth alone on November 17. On the way he picked

The Land between the Rivers

up two unnamed hitchhikers who sought passage to Lawrence-burgh, where they spent the night at a tavern. Nuttall the next morning was five miles downstream when he recalled where he had laid his gun at the tavern; it was too late to go back. No matter—Nuttall was neither a marksman nor a hunter and typically used his gun as a tool rather than a weapon. The winding river, descending toward Louisville, hosted a number of towns inhabited by Swiss, French, and Irish immigrants. One Frenchman was trying to coax from the soil vines that would make a good brandy, if not claret. Another "industrious Irishman, who had emigrated from Belfast" seventeen years before, opened his door and provided lodging for the fatigued traveler.

Nuttall spent two weeks at Louisville, Kentucky, awaiting passage to the Mississippi aboard one of the many steamboats that ascended the Ohio to the falls at Louisville. The river was low, however, and Nuttall impatient to proceed. He spent a fortnight exploring his environs, particularly the picturesque falls, the beauty and power of which increased as the level of the river decreased. At Louisville Nuttall saw the image and reality of the West, the idea that possessed so many Americans, "all searching for some better country, which ever lies to the west, as Eden did to the east." Nuttall found himself part of this "jarring vortex of heterogeneous population" who were all searching, most for wealth and opportunity, a few like Nuttall for knowledge.

Eventually the scientist decided to turn merchant and purchase a flatboat and "freight" her with supplies to be sold downstream. The cost for the goods was exorbitant, but Nuttall knew that he could recoup the amount downstream where scarcity would command even higher prices. He set out on December 7 accompanied by two men, a Mr. Godfrey and his son Edwin, who were on their way to New Orleans.

The Ohio from Louisville to the Mississippi grew broader, deeper, quicker, but with fewer hazards. The men made excellent time with little labor. The destitute inhabitants along the route paid dearly for salt and grain; Nuttall sold them supplies but condemned

The Ohio River, 1818

their indolence. Men chose to hunt rather than to farm, with predictable consequences for their families.

December 12 brought one of the finest days of Nuttall's journey. The river was wide and passive, the descent quick, the weather bright and wonderful. The night chill was even pleasant with the full moon smiling above, its reflection dancing upon the water. The banks of Indiana quickly went by; the mouth of the Wabash River heralded the state of Illinois.

From the Wabash to the mouth of the Ohio the river was smooth, the current fast, the navigation easy. Islands, the only real hazard, were easily avoided. The wooden, rectangular flatboat was equipped with several long oars, one of which was used as a rudder. The boat was not maneuverable but comfortable, with sturdy sides to keep the cargo dry and an enclosed, roofed area that served as a cabin. The boat was efficient on the lazy river. This holiday of sorts, the ease and splendor of the broad, friendly river, the clear sky dotted with moon and stars at night, the gentle rhythmic gurgling of the water, the relaxation of wilderness travel, came to an abrupt end at the mouth of the Ohio.

10 The Mississippi, Winter 1818

The confluence of the Ohio and Mississippi is 150 miles downriver from St. Louis. The region had been visited enough times by hunters, traders, missionaries, and explorers to be relatively well-known. Nuttall used a literary tour guide, as it were, Zadok Cramer's *The Navigator*, which included descriptions of varying worth and accuracy of the course and hazards of the Mississippi, as well as suggestions for navigating the river.

Nuttall and his companions were astonished to find ice floating on the Mississippi, the novelty of which forced them to make port to determine the best way to proceed. The "gloomy" forest extended from the river an unending distance. Fierce, "impenetrable" cane-brakes guarded the exit from the steep, muddy riverbank. Nuttall, of course, found a way to explore the environs. In a swamp dominated by the cypress (*Cupressus*), so prevalent along the Mississippi, he discovered milkweed (*Asclepias*), an essential ingredient in the Indian materia medica, and buttonbush (*Cephalanthus occidentalis*), a folk remedy for malaria. The coffee tree (*Gymnocladus canadensis*) produced a legume that made indifferent coffee but was otherwise good to eat.

Several flatboats having come to rest at the same spot, Nuttall conversed with their respective crews, some of whom were regular mariners on the Mississippi. They informed the scientist that "the whole country here, on both sides of the Mississippi and Ohio, remains uninhabited in consequence of inundation." Local hunters described the game to be had hereabouts—"deer and bear, turkeys, geese, and swans." Opportunities for the hunt attracted parties of Delaware and Shawnee Indians, one of whom Nuttall invited for dinner in the "cabin" of the flatboat. Nuttall, interested in expand-

The Mississippi, Winter 1818

ing his investigations into the customs and history of the Indians, sympathized with the plight of the Shawnees, driven before the advance of white civilization, eschewing settled farm communities for the nomadic ways of their forefathers, their numbers dwindling in the face of aggression from their enemies, the Cherokees and Osages. The Shawnee enjoyed the meal and conversed with his host by "the same symbolical or pantomimic language, as that which is employed by most of the nations with which I have become acquainted. It appears to be a compact invested by necessity, which gives that facility to communication denied to oral speech."

After a delay of two days, figuring that the ice on the river would only grow worse, Nuttall and his companions embarked on the broad Mississippi. Nuttall used his copy of *The Navigator* optimistically, hoping that the book would provide such a newcomer with moderately accurate information for navigating the Mississippi— identifying and numbering islands, dead ends, sunken forests, and other permanent hazards of the river. The guidebook proved generally useless, however, particularly when confronted with planters and sawyers, those recurrent and unexpected leviathans of the river. Nuttall discovered by experience the danger of the planter, a submerged tree driven by the current into one fixed spot, a deadly, immovable force should the unrelenting river propel the wooden flatboat into one. Sawyers, less dangerous, were moving logs, a collision with which could prove quite violent, even deadly.

The Mississippi demanded much more than the Ohio. The former required adaptable boatmen willing to become overnight experts in the art of navigating the maze of deep, rapid water and alternating shallows. Tremendous sand islands dammed the current, forcing Nuttall to decide instantly which side of an island appeared most navigable. The botanist wanted to observe the vegetation of the riverbanks, the vast stands of cane, and the scrubby island bushes. The geologist yearned to inspect the unstable shoreline, to read in the cliffs and extensive areas of sunken soil the recent as well as the distant past. The river revealed continual evidence of its cataclysmic history: steep shores showed signs of the current's constant dredging action. In one spot Nuttall saw unmistakable indications of the collapse of at least an acre of earth thirty to forty feet into the river. The flatboat repeatedly slipped in between gnarled earth; massive dams of driftwood; and sandbars, islands in the making.

One such "miry bar" of sand (labeled Wolf's Island by *The Navigator*) stopped the flatboat in midriver on the evening of December 19. Nuttall, "sensible of the dilemma into which we had fallen, . . . lost no time to plunge into the water, though at the point of freezing, attempting, but in vain, to float off the boat by a lever." The water was sufficiently shallow not to endanger Nuttall, who could not swim—though the water's frigidity was a different matter. Two passing boatmen, for a fee, used their flatboat to tow them off the bar and downriver. The next morning, the river having fallen overnight, the boat rested on yet another sandbar. Other boatmen assisted—again for a fee. Eventually, by late afternoon, Nuttall and his friends were on their way—unencumbered again, though thirteen dollars the poorer.

The sublime solitude of the wilderness astounded Nuttall and his companions. The river, "magnificent" yet unforgiving, dwarfed the men. Not a day passed without their seeing the remains of wrecked flatboats along the shore. Many a boatman had come this way down the Mississippi never to leave it. Grounded, overturned boats jutting from the water, the remains of flatboats trapped in gigantic masses of rotting trees and branches, served as headstones to remind the newcomer of the peril that awaited, perhaps around the next bend.

South of New Madrid, on the way to the mouth of the Arkansas River, were tales of woe written into the earth and heard on the lips of settlers about the great earthquake of the winter of 1811. Nuttall, who had passed this way on his first descent of the Mississippi, had luckily missed by only a few days being trapped by the quake. He could now, seven years later, see the toll it had taken. That cold December day whole towns had been lost and thousands of acres of land inundated. The shock had been felt over hundreds of thousands of square miles. The few inhabitants who remained in the

The Mississippi, Winter 1818

area were French-Canadians, destitute and indolent. At Little Prairie Nuttall learned that during the quake the land had sunk ten feet, opening it to the searching waters of the Mississippi.

The past was etched into the great elliptic bends of the river. Some bends were seven to eight miles long. The massive current of a river that is in places over a mile wide pounds the riverbank, lightly at the extremes of the bend, more forcefully at the center. Digging away over an incalculable amount of time, the bank gives way to the river and a new channel is formed. What had been the furthest extreme of the bend is cut off, forming a massive sandbar, soon to be an island.

Observing the natural record of the past, Nuttall could sense the erratic rhythm of the river—never stagnant, always moving, shifting, creating new paths, erasing old ones, sometimes slow and ambling, often a powerful destructive force, rarely gentle. The river was an excavator of future paths even as it directed the fleeting present and refused to silence the echoes of the past.

The isolated inhabitants who lived along the banks of the Mississippi were, like the river journeyers, simultaneously attracted to and repelled by the river. The river was the reluctant host of visitors and residents alike, as well as a recalcitrant conveyor of dreams into reality. The squatters of the region were miserably poor farmers and hunters dressed in buckskin, wearing moccasins. Generally Nuttall saw only reminders—abandoned shells of houses, destroyed flatboats, fallen riverbanks, sunken forests.

On December 29, approaching the Chickasaw Bluffs of Tennessee, Nuttall heard the distant sound of rapids. The boat soon approached a terrifying stretch of river that Nuttall likened to the impossible passage between Scylla and Charybdis. The river narrowed, and the current quickened, gushing and roaring amid hundreds of planters barring the stream. Adding to the scenario of apparent doom was a downpour of torrential rain, the scattered wrecks of other flatboats where mariners had breathed their last, and a landscape of "houseless solitude and gloomy silence." The men felt overwhelmingly alone, yet they had little time for despair

The Land between the Rivers

as they fought for their lives in this "labyrinth of danger," a river "horribly filled with black and gigantic trunks of trees." Miraculously, the flatboat bumped and slid its way past disaster to safety.

Near a place called Flour Island they found several families of destitute hunters, including some members of the Delaware and Shawnee tribes. The Indians had little to eat but many beaver and muskrat pelts, which they "were anxious to barter . . . for whiskey." On New Year's Day 1819 Nuttall and companions arrived at a "Shawnee camp," the residents of which were interested in the same commodity; Nuttall "bartered for some venison and wild honey, which they had in plenty. The honey, according to the Indian mode, was contained in the skin of a deer taken off by the aperture of the neck, thus answering, though very rudely, the purpose of a bottle."

The first week of January the men navigated river hazards termed by the locals the "Devil's Race-ground" and the "Devil's Elbow," neither of which seriously threatened the flatboat, though the visible remains of wrecked boats were everywhere. On January 4 Nuttall arrived at Fort Pickering (now Memphis), where there was a government store that sold supplies to settlers, Indians, hunters, trappers, and ne'er-do-wells. Whiskey was sold, bought, and drunk freely, to the detriment of all. This was the land of the Chickasaw tribe, the members of which generally impressed Nuttall by their dress, deportment, and knowledge of English, notwithstanding the apparent intoxication of scores of men, women, and children.

Several days downriver the boatmen arrived at other, similar settlements. The locals hungered for whiskey and lived miserable lives, their tales of woe centering on the memory of the destruction of the 1811 New Madrid earthquake, which had annihilated thriving settlements and sent many to their doom. Nuttall took time to wander several miles into the forest, noting the variety of vegetation, particularly the massive, leafless cottonwoods, "some of them more than six feet in diameter, and occasionally festooned with the largest vines which I ever beheld." Mistletoe grew on the hackberry (*Celtis occidentalis*), the hornbeam (*Carpinus caroliniana*), and the beech.

The Mississippi, Winter 1818

The forest featured holly (*Ilex opaca*), useful in treating the symptoms of malaria; horsetail (*Equisetum*), a folk remedy for urinary problems; and buttonbush (*Cephalanthus occidentalis*), on "the seeds of which flocks of screaming parrots were greedily feeding." Cane (*Arundinaria*) was ubiquitous, as were different species of the willow (*Salix*).

The next few days of navigation presented further obstructions, signs of a cataclysmic natural past, and the unending desolate forest. The number of flatboat graveyards, remains of boats caught in and henceforth irrevocably linked to the massive dams of driftwood and trees from disemboweled lands upstream, was sobering. Nuttall, by now an experienced boatman, knew to avoid the rushing current that drew together piles of driftwood. Instead, the navigator had to stay to the opposite shore, where the current was weaker and the sandbars more profuse. Vines and briars shrouded the banks of the Mississippi. Greenbrier (*Smilax*) and supplejack (*Berchemia scandens*) were dominant, the former being, like its cousin the sarsaparilla (*Aralia nudicaulis*), a particularly effective cure-all. Several species of honey locust (*Gleditsia*) grew nearby. Along the way Nuttall met with a band of Quapaw Indians seeking whiskey, which he reluctantly sold to them. Near Island #66 (according to the itinerary of *The Navigator*) the combination of warm air and icy water spawned a dense fog that made the already scarcely detectable planters and sawyers even more terrifying. Yet the three men, Nuttall, Mr. Godfrey, and his son, had by experience gained considerable ability to guide the flatboat to safety.

Nuttall parted from his companions, and the Mississippi, on January 12, upon the advice of Neil McLane, who owned a tavern at the mouth of the White River. McLane told Nuttall that the best way to reach the Arkansas River was by means of the White River, which was swollen with backwater from the Mississippi. To help him pole the flatboat upstream, Nuttall hired a "drunken . . . Yankee" who demanded a salary of ten dollars, as well as the flatboat and cargo upon their supposed arrival together at the Arkansas River. Waiting for the Yankee to sober up, Nuttall strolled about the canebrakes at the river's mouth. He found a species of ragwort (*Senecio aureus*), which, because Indians used it to treat diseases peculiar to females, was known as squaw-weed. Nuttall soon discovered his employee to be a "worthless" scoundrel who attempted, apparently, to rob his employer, forcing him to hire another in his place. This unnamed adventurer and Nuttall towed and tugged the flatboat through the shallow "bayou" of the White River. Their progress was slow, the effort exhausting, the surrounding swamp an uninhabited, cheerless wilderness. Eventually, on January 16, 1819, Nuttall and his companion reached the "red and turbid" Arkansas River.

Such was Thomas Nuttall's introduction to the Arkansas valley, where he would spend the next eleven months attempting to ascend the river to its source in the Rocky Mountains, thereby fulfilling his dream to explore as a scientist and natural historian the dominant river of the southern territories of the United States of America.

11 The Arkansas, Winter 1819

The Arkansas at its confluence with the White River was several hundred yards in breadth, though much of that was barren sand. The channels of water flowing southeast were not deep. The men became quickly exhausted towing the flatboat "by means of the cordelle, but at a very tedious and tiresome rate." They towed alternately on the bank or in the water, wading for hours in the cold river. January along the Arkansas was nevertheless not entirely unpleasant. Temperatures during the day reached into the sixties. The sun was distant on the southern horizon, yet warm and bright. Nuttall observed cane growing at the top of steep cliffs on the side of river bends where the current was most fierce; sandbars, poplars, and willows dominated the opposite side. Nuttall, not yet introduced to quicksand, objectively noted that "the sand of the river appears to be in perpetual motion, drifting along at the beck of the current." This region of the river, near its mouth, was continually subject to inundations from the Arkansas and especially the Mississippi, the backflow of which during floods could drown thousands of square miles. The past of these changing waters marked the region with a dismal solemnity, where all was silent, apparently lifeless save for the barren trees and a few hearty birds. Human voices rarely echoed in and about the sand and cane. Nuttall believed the land still bore the marks of its primeval beginnings. This was Nuttall's dream, to journey into virgin territory, to be the first botanist to ascend the Arkansas; the thrill overcame his misgivings, utter fatigue, and the apparent danger of the land. Timothy Flint, who traveled through Arkansas about the same time, called the region New Kentucky and claimed that the hunters who lived there were savage and primitive. More people from the civilized East were

The Land between the Rivers

indeed moving to this New Kentucky, which was still, however, in 1819 a merciless land of the Indian, trapper, hunter, voyageur, and ne'er-do-well.

At the point where the travelers reached the first tavern along the Arkansas the arboreal environment changed. A delighted Nuttall journeyed along the banks of the river, glorying in the pecan (*Carya illinoensis*), at this season sans its nuts; the black walnut (*Juglans nigra*), suggestive of dozens of folk remedies for ailments ranging from toothache to cancer; stands of cottonwood (*Populus deltoides*), honey locust (*Gleditsia*), and poplar (*Populus*); and the varied forest of the *Quercus:* swamp oak, red oak, scarlet oak, Spanish oak. The ubiquitous swamp willow (*Salix nigra*) drooped and wept, but without the verdure of its fingerlike leaves. The bark of the sycamore (*Platanus occidentalis*) peeled, as usual, suggesting in Nuttall's mind its traditional efficacy against rheumatism and the scurvy. The trees were leafless, of course, and strangely stunted in size; Nuttall theorized that the sandy soil and unstable banks supported insignificant growth in trees.

The scientist, realizing the futility of poling his bulky flatboat up the Arkansas, approached some merchants taking trade goods upriver the 350 miles to Fort Smith, his initial destination. The scientist hoped for some indulgence, presenting his credentials and interests to the merchants. But they "appeared to be illiterate men, and of course, . . . incapable of appreciating the value of science." Waiting for another means of travel, Nuttall discovered there was a land route to the Arkansas Post, which was 15 miles upriver. The trail took him through a "horrid" oak swamp to a vast prairie, in which there were already wildflowers abloom. Perhaps he munched on the leaves and flowers of the blue violet (*Viola sororia*). He also saw a species of bluet (*Houstonia*) in bloom. The cardinal (*Cardinalis cardinalis*) chirped happily, delighting in the springlike weather. Nuttall wrote years later that "the notes of our Cardinal are as full of hilarity as of tender expression; his whistling call is uttered in the broad glare of day, and is heard predominant over most of the feathered choir by which he is surrounded. His responding mate is

The Arkansas, Winter 1819

the perpetual companion of all his joys and cares; simple and content in his attachment, he is a stranger to capricious romance of feeling; and the shades of melancholy, however feeble and transient, find no harbour in his preoccupied affections." Also beautiful to look at and listen to was the bluebird (*Sialia sialis*). Because the bluebird, Nuttall wrote in his *Ornithology*, is "gentle, peaceable, and familiar, when undisturbed, his society is courted by every lover of rural scenery."

Arriving at Arkansas Post, also called the Town of Arkansas, the first settlement of the French on the Arkansas and capital of what was about to become the Arkansas Territory, Nuttall called on Joseph Bogy and presented his letter of introduction from Colonel Hugh Glenn. Bogy was seventy years old but very fit and active. Nuttall perceived in him the manners of a gentleman although he was dressed "in the garb of a Canadian boatman." Bogy agreed to help the young scientist. His plans set, Nuttall returned to his flatboat on the same path, the swamp now swelled to knee-deep, stagnant water. Nuttall did not see any alligators, but he knew from hearsay that they inhabited the swamp.

On the morning of January 24, Nuttall proceeded the few miles upriver to Arkansas Post, assisted by two Canadian "boatmen, full of talk, and . . . indifferently inclined to work." Upon arrival, Nuttall sold his boat and remaining stock, then patiently awaited an opportunity to proceed toward his goal. The scientist spent several pleasant days at Arkansas Post, exploring, talking with the inhabitants, learning the history of the region, and writing letters. He learned about the Arkansas valley and the Three Forks from Bogy, who suggested that Nuttall alter his plans. Nuttall was agreeable, as revealed in a letter to his friend Zaccheus Collins. Nuttall wrote that he now intended to find passage to Fort Smith, from which point he hoped "to proceed up the Canadian river to the mountains either by land or by water as may be most convenient, and afterwards to descend Red river." The Canadian River is the first important western river flowing into the Arkansas River west of Fort Smith. Hunters claimed to have followed this river to the foothills of the Rockies. And this plan had the added convenience of avoiding the Osage Indians at the Three Forks and beyond on the Arkansas River.

Also at Arkansas Post Nuttall met and conversed and botanized with Dr. McKay. The doctor informed the scientist of the habits and inclinations of his French neighbors. The French rarely visited a physician, owing to their belief in traditional remedies. They appeared indolent and too fond of frivolity, as evident in the poverty of their farms and towns. French-Canadian explorers of the seventeenth century who ascended the Arkansas under La Salle to Arkansas Post had stayed, intermarried with the local Quapaws, and adopted the Indian lifestyle of hunting and casual farming. Nuttall despised the descendants of these French-Indians and rather wished that "9000 Germans from the Palatinate" had immigrated here, as "we should now probably have witnessed an extensive and flourishing colony in place of a wilderness, still struggling with all the privations of savage life."

Nuttall accompanied Dr. McKay on excursions into the surrounding prairies, delightfully decorated by stands of a species of spring beauty (*Claytonia virginica*), a delicate purple and white flower, the root of which was good tasting. There was also a profusion of bluets (*Houstonia caerulea*). A physician and a botanist on a stroll could not resist, of course, adding to their pharmaceutical knowledge. They saw the aromatic sweet gum (*Liquidambar styraciflua*) and perhaps obtained the gummy resin underneath its bark; the gum was a standard folk remedy as well as an interesting chewing experience. They also found the white-flowered rattlesnake master (*Eryngium yuccifolium*), used by the Indians for snakebites. Dr. McKay told Nuttall the French locals took it "as a diuretic." Box elder (*Acer negundo*) grew here. The southern prickly ash (*Zanthoxylum clava-herculis*), its bark a sure cure for toothache, was in abundance. So, too, were abnormally large rats. But countering such ugliness was the beauty of fruit trees, already in bloom.

Eager to continue his journey into natural history, Nuttall joined a group of traders and travelers ascending the river some thirty-odd miles. They camped on the vast sandbars of the Arkansas at night to avoid mosquitoes. Although some of the journeyers complained about their sandy accommodations, Nuttall proclaimed himself satisfied as long as the days were warm, the evenings clear, and the stomach full. Settlements varied along the river shore. The north side had sporadic settlements of three to four families trying to scratch out a miserable existence. The south side had similar settlements of Quapaw (Arkansas) Indians engaged in the same, often fruitless task. Nuttall had long been fascinated by these people, having read about their courage and dignity in the works of two sympathetic French explorers and writers: Le Page du Pratz, author of *History of Louisiana* (1763), and Father Pierre Charlevoix, author of *Journal of a Voyage to North America* (1761). Nuttall considered the Quapaw a well-proportioned, handsome people. The men wore the typical apparel of the frontier: moccasins, leggings reaching to the thigh, a breechcloth covering bare loins, and a hunting shirt. They were excellent warriors and successful hunters but reluctant husbandmen. They referred to themselves as O-huah-pa; the French called them Osarks, and the river upon which they lived the Rivière des Arks or the Rivière d'Osark.

As the large pirogue ascended the river, Nuttall did what he could to escape his resting comrades for botanical walks. On the first day of March he followed a path into "a horrid cane-brake, interlaced with brambles, through which I had to make my way as it were by inches." Such labor scarcely cooled his desire for such excursions. On another jaunt a week later Nuttall journeyed through woods the forest floor of which was decorated with the buttercup (*Ranunculus acris*), beautiful to look at, like the rose, but with its own type of thorns—acrid leaves. At the home and tavern of Toussaint Dardenne, Nuttall rendezvoused with William Drope, a merchant whom Nuttall had met at Arkansas Post. Drope was ascending the Arkansas in a large flatboat loaded with trade goods. Nuttall hitched a ride. They made slow progress, fifteen to twenty miles a day at the most, halting at night at settlers' homes or camping on immense sandbars. The beautiful (and edible) flowers of the redbud (*Cercis canadensis*) decorated the budding forest of cottonwood, maple, ash, sycamore, and elm that lined the shore. They saw as well the holly (*Ilex opaca*), which locals used as a laxative and otherwise to purge the system, a common response to illness at the time. Plum (*Prunus americana*) was in flower, though it would be some time before the appearance of the fruit. Cypress (*Cupressus disticha*) emerged from some of the thickets of cane. The vines of wild grapes (*Vitis*) wound about the forest, the fruits, seeds, and leaves of which had an important role in the Indian materia medica.

By the time they approached the Little Rock, in mid-March, winter again showed itself with cold winds and comparable temperatures. Ice, which Nuttall had not yet encountered on the river, began to form in places. Flowering plants, taunted with hints of spring, again faced the reality of winter. The vicinity of the Little Rock featured imposing bluffs decorated with pine trees rising from the riverbank. Nuttall learned of ancient Indian paths that traversed the region, one heading to the Red River, over two hundred miles away, another to the Hot Springs, fifty miles away. The surrounding land of forest and prairie grew more hilly with each day of travel. Nuttall's excitement grew as he found himself again embraced by mountains, however small, such as the Mamelle. Mountains host gems and rich ore, as all the locals and hunters knew; they informed Nuttall of a silver mine near White Oak Bayou. Nuttall, duly suspicious, journeyed to the region, which was beautiful enough, exhibiting shiny rocks. Nuttall informed the incredulous locals, however, that "the microscope" (which he carried for such occasions) did not "betray the smallest metallic vestige which could be taken for silver." Silver mines, long talked about by the French, Nuttall branded "a fairy dream." On March 23 Nuttall ascended one of the local peaks to survey the Arkansas valley and surrounding territory. He saw small ranges of mountains extending northwest and southwest, some of them quite beautiful. The peak of the Mamelle was chastely hidden in the clouds. Nuttall took time to sketch the mountain and valley.

Toward the end of March they arrived at the Cadron settlement, a crossroads and a pastoral community surrounded by romantic

The Arkansas, Winter 1819

cliffs and hills. Nuttall, restless, again explored the environs, disappointed with the lack of industry among the inhabitants, wondering if and when civilization would overtake savagery among the American immigrants and descendants of French voyageurs. Even so, Nuttall was surprised to see how healthy the inhabitants were, how scarce was the debility from which he often suffered, the ague. The spicebush (*Lindera benzoin*) was in profusion, its tea a healthy counter to a variety of illnesses such as arthritis and colic. Agriculture tended to be neglected in favor of the hunt. Bison ranged the nearby prairies, and the bobcat (*Felis rufus*) populated the woods. The songs of birds and trill of the river were sometimes interrupted by the noise of the grist- and sawmill recently built by John Benedict.

The first week of April, the weather still more like winter than spring, the journeyers passed the mouth of the Petit John River, which descends from mountains to the west. Nuttall took time to journey and climb small hills. Some were very rocky, the stones delicately balanced one upon another as if set by a great builder. Here Nuttall discovered a species of the delicate windflower (*Thalictrum thalictroides*). The Arkansas River, due to spring rains and melting snow hundreds of miles away, continually rose, and its current quickened. Locals informed Nuttall that the river's rising would peak in May, which led Nuttall to deduce that "no considerable branch of this river derives its source within the region of perpetual snow, which dissolves most in the warmest season of the year."

At a place called the Dardanelle Nuttall rambled about the hills and forests, delighting in the sights and sounds that greeted his eyes and ears. The dogwood (*Cornus florida*), its white flowers showy and beautiful, lifted the spirits. Dusk brought the sound of crickets chirping, frogs singing, and the enchanting notes of the whippoorwill (*Caprimulgus vociferus*). On one ramble Nuttall spied what he thought was a mole retreating under a rock, which when lifted revealed a huge spider, the tarantula (Theraphosidae). More disconcerting, perhaps, were the ticks (*Acarus sanguisugas*), fifty of which Nuttall "picked off my skin and clothes."

The Land between the Rivers

At Dardanelle Nuttall again saw Reuben Lewis, his companion during his descent of the Missouri in 1811. Lewis informed Nuttall of the state of the Cherokees, who populated this region of the Arkansas. They were recent immigrants, having taken over the land from the Osage Indians in 1808 because of a treaty negotiated by William Clark. Yet hard feelings continued between the two rival tribes. Nuttall discovered the Cherokees moving happily "towards civilization." They cultivated the soil and built farms and fed their children better than most of their French and American neighbors. On April 10 Nuttall met Walter Webber and Tallantusky, leaders of the Cherokees. They lived and dressed in style, directed the active husbandry of their people, owned slaves, and were known as thinkers among their people. Nuttall thought Tallantusky, the chief, remarkably cultivated and civilized, notwithstanding his bloodthirsty reputation among Osages who had survived the Cherokee attack of 1817. Indeed, Nuttall wondered whether or not a people who pursued aggressive war against others, including the aged, infirm, and children, could be counted among the civilized. He also condemned their constant blood vengeance, superstitions, and treatment of women, who were but slaves.

William Drope, having exhausted his trade goods, halted at Webber's farm and prepared to descend the Arkansas again, which left Nuttall without the means of ascending the river to Fort Smith. Fortunately he found two French voyageurs who owned a pirogue and who were willing to take him the short distance to the fort. They passed through a region of low hills and plentiful game. Nuttall, at rest on the voyage, examined the landscape of the Arkansas valley: "The beauty of the scenery was . . . enlivened by the melody of innumerable birds, and the gentle humming of wild bees, feeding on the early blooming willows" (*Salix caroliniana*). Cottonwoods hovered about the banks of the river, their delicate leaves having emerged plentifully under the now warm sun of late April; even the gentlest zephyr rustled the dark green leaves. Competing for dominance of the flood plain was the towering black (swamp) willow (*S. nigra*), its pincerlike leaves pointing in the direction of the wind. Nuttall

The Arkansas, Winter 1819

thought the Arkansas resembled the Ohio, although the water of the latter was much more clear. Beyond the river valley, like that of the Ohio, were extensive plains, though the plains of the Arkansas Territory had few trees. What trees there were "clad" the river valley "in the softest and most vivid verdure." Animals and birds fled the burning sun of the plains for the cool water and shade of the river. Nuttall and the Canadians burned under the sun, too; the thermometer showed it to be 100 degrees. The cool air of evening challenged the still radiant afternoon heat to produce copious clouds, thunder, and storms. On April 24 the river made a huge bend to the south, passing by Lee's Creek to the north. At noon they reached the confluence of the clear and rapid Poteau River, rushing out of hills to the south, and the Arkansas, still sandy and slow. Here, on a bluff overlooking the two rivers, the United States had erected its westernmost fort in the former Louisiana Territory. Major William Bradford had received the assignment, chosen a place called Belle Point by the French, and renamed the army post Fort Smith.

The soldiers under Major Bradford had grown accustomed to the comings and goings of many strange characters. Numerous travelers and desperadoes passed through Fort Smith, ascending the Arkansas and the Poteau to hunt, trap, trade, or even escape their past. The soldiers hardly expected to see the likes of the law-abiding, seed-gathering, absentminded, transplanted Englishman Thomas Nuttall. This frontier, on the edge of civilization, on the fringe of law and order, was certainly not a place for the foolhardy.

12 Belle Point, 1819

Thomas Nuttall was not the first scientist to ascend the Arkansas River. Preceding his coming were explorers who became ad hoc scientists, soldiers experienced in the arts of surveying and cartography, and physicians. Medical science in the early nineteenth century was still sufficiently precarious, based on folklore and home remedies. William Smith's complaint in 1757 regarding medicine in America that "quacks abound like locusts in Egypt" still held true in large part fifty years later. Most Americans, if they worried at all about health and practiced *physic,* relied on books such as John Tennent's *Everyman His Own Doctor: Or, the Poor Planter's Physician* (1734) or, for women, *The Compleat Housewife; Or, Accomplished Gentlewoman's Companion* (1742). A few physicians had training from European universities, such as the University of Edinburgh. But even these trained physicians continued to be indebted to the great classical medical writers Hippocrates and Galen. Medical studies of the Renaissance and Enlightenment (1500–1800) had challenged some nonsense from the past, such as the notion that an imbalance in the four humours (blood, phlegm, black bile, yellow bile) was the general cause of illness. Human anatomy and the circulatory system were better understood thanks to the labors of Andreas Vesalius and William Harvey. If microorganisms, bacteria, and viruses were not yet understood in 1800, early American scientists were increasingly aware of the importance of the environment, diet and sanitation, in causing disease. Yet so able a physician as Benjamin Rush, who was a pioneer in chemistry, psychology, dentistry, and ideas on the causes of disease, was still beholden to the archaic notion of bleeding those who were ill, particularly with high fever, which was a throwback to the ideas of Hippocrates. Benjamin Rush was before

Belle Point, 1819

his death in 1813 a professor of medicine at the University of Pennsylvania, which had one of the few American medical schools. Upon Rush's death Benjamin Smith Barton assumed the professorship in "the Theory and Practice of Medicine." Among Barton's students, and before him Rush's, was a young man from Salem, Massachusetts, Thomas Russell.

Thomas Russell graduated with the M.D. from the University of Pennsylvania in 1814. Thereafter the unmarried doctor joined the U.S. Army as a hospital surgeon. After Major Stephen Long had selected the site of Belle Point as the location of a new fort, and upon the assignment of troops to the new Fort Smith in 1818, Thomas Russell became post surgeon—the army physician stationed at the fort. Unlike today, two centuries ago good physicians were almost always good botanists. Thomas Nuttall was therefore extremely delighted to find, upon his arrival at Fort Smith at the end of April 1819, Dr. Thomas Russell. The two immediately realized their common interests and personalities, which separated them from the hunters, soldiers, and voyageurs who lived and worked in Fort Smith and its vicinity; they formed a close, if brief, friendship.

Army soldiers had selected a splendid spot to build the fort. The waters at the confluence of the Poteau and Arkansas were wide and deep, coming together in such a way as to form a triangular peninsula. Nuttall thought the view from the fort "is more commanding and picturesque, than any other spot of equal elevation on the banks of the Arkansa." The fort, which consisted "of two blockhouses, and lines of cabins or barracks for the accommodation of 70 men whom it contains, is agreeably situated . . . on a rising ground of about 50 feet elevation, and surrounded by alluvial and uplands of unusual fertility."

For three weeks Nuttall enjoyed the hospitality of the fort. He went on almost daily excursions into the surrounding countryside and ranted and raved about the prairies, abloom in spring glory. Every so often Nuttall spied in the distance, in the direction of the Poteau River, "a conic mountain nearly as blue as the sky, and

97

The Land between the Rivers

known by the French hunters under the name, Point de Sucre, or the sugar loaf." Adjacent to it was the "tabular summit" of Cavaniol Mountain, which sufficiently impressed Nuttall that he sketched it. He was attracted to its massive immovable silence and "to the sublimity of the clouded mountain." Standing apart from the rest, it appeared majestic and eternal, "contrasted with that transient scene in which we ourselves only appear to act a momentary part."

Also exciting to the botanist were the prairies surrounding Fort Smith: "The whole expanse of forest, hill, and dale, was now richly enamelled with a profusion of beautiful and curious flowers." Nuttall discovered the "American Daisy" (*Bellis*), a smaller counterpart to the daisy of European origin; the spiderwort (*Tradescantia virginiana*), useful for treating spider bites; a species of vervain (*Verbena*), a cure-all for digestive complaints; Osage orange (*Maclura pomifera*), useful for dye; primrose (*Primula*), dazzling now in the spring sun; and the wonderful Indian paintbrush (*Castilleja coccinea*), which the Indians reputedly used to encourage love (as an aphrodisiac) or bring on death (as a poison). On May 3 Nuttall and Dr. Russell "rode to Cedar prairie, lying about 10 miles south-east of the garrison." The two found growing the soft yellow flowers of cream wild indigo (*Baptisia leucophaea*), which local Indians used for colic and fever; the purple-flowered blue-eyed grass (*Sisyrinchium anceps*); and the rich blue larkspur (*Delphinium*), named for the lark. A week later the doctor and botanist again visited Cedar Prairie, where Nuttall discovered the sublime silence of nature, with the wind causing the only sound. After sundown frogs, coyotes, and whippoorwills broke the silence with the sounds of night.

13 Kiamichi Wilderness, Spring 1819

They moved southwest through prairie grass that reached a man's waist. The morning dew wetted their buckskin and wool pantaloons. The hunters and guides assured the rest that they knew the way, even without well-marked paths, much less roads. Enough deer came this way to forge a narrow trace. The disbelief of the others was, perhaps, dispelled by the frequent deer that leaped from their hidden abodes at the coming of the men. Indeed, the prairies abounded with wildlife. The sporadic groves of weathered cottonwoods indicating small creeks hid their share of mysteries and wonders. One hunter, the pilot of the expedition, found a tree storing the golden fruit of the honeybee. The men's sticky repast included the typical epicureans of the forest. Two of these bears, fattened by honey, berries, and insects, were discovered and killed by the hunters. Bear meat and honey made for an ironic meal. Even so, the repast was scarcely enough to feed the numbers of hunters and soldiers who forged a path through the wilderness of the Arkansas Territory on their way from Fort Smith southwest through the Kiamichi Mountains to the Red River. Such a journey was hazardous and difficult, which is always the case when for weeks one must live off the land, find the way amid the limitless chaos of nature, the unending prairie grass, clumps of trees, shallow muddy creeks, impenetrable thickets. The wind, dust, flies, mosquitoes, rattlesnakes, ticks, bobcats, bears, thornbushes, sudden storms, heat of the noonday sun, and cool of the morning mist made for a plentiful array of dangers and sufferings that could occupy the minds and bodies of the most intrepid frontiersmen.

These were hard men, to be sure, soldiers and hunters brought to the edge of civilization, wading through the vastness of the wilder-

The Land between the Rivers

ness, joined by a common longing for something else, something more. Among them was an anomaly, an exception to the standard type of man one found in the Arkansas Territory of 1819. If Thomas Nuttall possessed the same yearning to survive and to avoid danger, nevertheless he saw amid the trials and privations of the daily struggle through the forests, mountains, prairies, and rivulets an unsurpassed beauty scarcely imaginable except to the one who sees it. "Our route," he wrote in his journal, "was continued through prairies, occasionally divided by somber belts of timber, which serve to mark the course of the rivulets. These vast plains, beautiful almost as the fancied Elysium, were now enameled with innumerable flowers, among the most splendid of which were the azure Larkspur, gilded Coreopsides, Rudbeckias, fragrant Phloxes, and the purple Psilotria. Serene and charming as the blissful regions of fancy, nothing here appeared to exist but what contributes to harmony." Beautiful, perhaps, the hunter might say—but serene? charming? What fool would mistake the chaotic happenstance of the wilderness for charm and serenity?

Nuttall had responded to Major William Bradford's invitation to accompany a half dozen soldiers and a few guides and hunters on a journey to the Red River. There was no set trail, so the guides forged one as they went. Their object was to remove the squatters who had illegally taken and farmed the land of the Red River valley, which the government in Washington had decided should be reserved for the Osages, later the Choctaws.

For the first few days the wildflowers and grasses of the prairie and the unexpected Cavaniol and Sugar Loaf rising from the horizon distracted Nuttall. The region reminded him of the Allegheny Mountains of Pennsylvania, though these mountains of the Arkansas Territory appeared less wooded because the drier climate stunted the trees. The eastern red cedar (*Juniperus virginiana*), which will grow just about anywhere, grew here, as did a variety of oaks and hickories. The Winding Stair Mountains, which formed the watershed separating the Poteau and Kiamichi Rivers, were

Kiamichi Wilderness, Spring 1819

(and still are) a perfect wilderness of seemingly impenetrable forest with vines, brush, and brambles covering the forest floor.

Naturalists call the Ouachita Mountains an oak-hickory forest, and indeed the forest abounds with delightful arboreal variety. Compared with the forests of the highest peaks of the Appalachians in the northeast and Canada—dominated by conifers; massive ruminators such as the moose; raptors such as the great gray owl; and in some places, such as Mount Washington, a frigid alpine environment—the oak-hickory forest is quite different. The climate is more arid, though not as much as just a hundred or so miles to the west. Reduced humidity and rainfall, compared to the Appalachian Mountains and the surrounding forests of Maine, New Hampshire, Vermont, and New York west to the Great Lakes, mean that trees achieve less stature. Nuttall never saw in the Kiamichi wilderness the giant cottonwoods he had seen in the Missouri River valley or the white pine (*Pinus strobus*) that dominated the Appalachian lowlands. In the Arkansas Territory cottonwoods were abundant, if reduced in size. The canopy of the forest did not soar so high, and branches spread all the way to the forest floor, which was thick with undergrowth and the deciduous leaves of the past autumn and others more distant in time. The soft forest floor of the Appalachian wilderness differs from that of the oak-hickory forest, where there is little of the spongy ground caused by countless bogs, and the limestone foundation of this forest does not give very much. As Nuttall and the soldiers ascended the small peaks—none over three thousand feet—of the Kiamichi wilderness, there was no change in the undergrowth or trees; the forest was deciduous throughout.

There were dramatic changes depending on soil and elevation, however, according to relative moisture. Oaks and hickories were ubiquitous in this diverse environment. Dry, rocky ridges hosted the Ozark chestnut (*Castanea ozarkensis*), with small edible nuts; Spanish oak (*Quercus falcata*); bluejack oak (*Q. incana*), one of a few oaks with singular oblong rather than lobed leaves; bur oak (*Q. macrocarpa*); blackjack oak (*Q. marilandica*); chinkapin oak (*Q. muehlen-*

The Land between the Rivers

bergii), named for the Pennsylvania botanist Gotthilf Muehlenberg; post oak (*Q. stellata*); and black oak (*Q. velutina*), its yellow inner bark used by early Americans for dye. Varieties of hickories in the highlands included the pignut hickory (*Carya glabra*), the nuts of which are edible, if preferred more by swine than humans, and the mockernut hickory (*C. tomentosa*). In the moist lowlands of creeks and ponds the explorers saw the white oak (*Q. alba*), thought to be a remedy for skin and intestinal ailments; water oak (*Q. nigra*), found especially around swamps; pin oak (*Q. palustris*); willow oak (*Q. phellos*), its oblong leaves resembling its namesake; red oak (*Q. rubra*); and swamp oak (*Q. shumardii*). The pecan (*C. illinoensis*) was bountiful, its nuts prized by settlers and Indians alike. Valleys also hosted the shagbark hickory (*C. ovata*) and bitternut hickory (*C. cordiformis*), the oil from its nuts believed to be a salve for rheumatism. The aromatic sycamore (*Platanus occidentalis*) was ubiquitous in well-watered places, its broad canopy of large green leaves providing frequent, much-needed shade on these late spring days.

Oaks in particular have an extensive root system, which is needed to gather what little moisture exists during the summer heat waves of these parts. In many places west of the Kiamichi the oaks and hickories grew less numerous, replaced by the grasses of the prairie. Bison fed off the grasses and sometimes wandered east into the foothills and valleys of the mountains. On one occasion, as the men were descending a rocky ridge, they spied bison loitering in the prairie, but they started and fled with the soldiers in pursuit. Nuttall watched as "the bulls, now lean and agile, galloped along the plain with prodigious swiftness, like so many huge lions. The pendant beard, large head hid in bushy locks, with the rest of the body nearly divested of hair, give a peculiar and characteristic grace to this animal when in motion." The hunters told Nuttall that "the male, infuriate and jealous in his amours, gores every thing which falls in his way, and becomes totally unmanageable."

The nuts of the oaks and hickories provided plentiful food for birds and rodents, which in turn attracted predators such as the infrequently seen panther, or mountain lion (*Felis concolor*). Nuttall

102

Kiamichi Wilderness, Spring 1819

had to rely on the hunters for information, which tended to be anecdotal. For instance, he heard of one panther that simultaneously killed a hunting dog, deer, and wolf. Likewise the stories of the black bear (*Ursus americanus*) were astonishing—how they ate their prey before killing it and were so gluttonous about deer that hunters could imitate the bleat of a fawn and "decoy" the bear "within gunshot." Nuttall knew from his own studies that the bear's diet is usually more humble: "the common American species feeds upon fruits, honey, wasps, and bees; they will turn over large logs in quest of other insects."

Most of the animal inhabitants of this mountainous country were neither seen nor heard, but they were nevertheless present. Rodents such as the white-footed mouse (*Peromyscus leucopus*) and deer mouse (*P. maniculatus*) were less obvious than the eastern chipmunk (*Tamias striatus*), a spunky creature that emerged from the undergrowth to brave the ascent of any tree that promised acorns and nuts. The fox squirrel (*Sciurus niger*) particularly liked the oak-hickory forest; its cousin the gray squirrel was the most common presence in the forest, scurrying up oaks, taunting predators, even scolding at times. The cautious eastern cottontail (*Sylvilagus floridanus*) and the fat, seemingly content woodchuck (*Marmota monax*) could also be seen during the day. Even the red fox (*Vulpes vulpes*) sometimes betrayed its presence, being surprised on the trail or seen by the early-morning riser. The cry of the fox just as the faint light of dawn appears sounds eerie indeed, rather like that of an injured goose or alarmed crane. Likewise Nuttall and the men nightly heard the cry of the coyote (*Canis latranus*), though in the darkness of the night it sounded (at least to the imagination) like that of the wolf (*C. lupus*).

The night of May 19 it rained, which delayed their start the next morning and caused further trouble for the remainder of the day because the typically dry gullies suddenly hosted torrents of runoff. At the confluence of the Kiamichi with Rock Creek the river rushed through "very lofty ridges, partly covered with pine and oak." The bison trace they had followed all day disappeared, and the guides

The Land between the Rivers

were left to forge the trail alone. The region reminded Nuttall of the Blue Ridge Mountains of Virginia. The next morning he found the wilderness trail to be "equal in difficulty to any in the Alleghany [sic] Mountains." There was nothing gentle or easy about this river valley. Paralleling it soon became impossible because of the deepening gorge carved by the Kiamichi. The guides' suggestion that they find new bearings was hardly better. The men scattered through thick woods filled with ticks. Arriving at an insignificant creek with significant and challenging ridges, the guides informed the exasperated Nuttall, in response to his query, that *most* of the rivers of the Arkansas Territory were unlike the Kiamichi; rather, they had broad floodplains and lazy paths through sparsely wooded landscapes.

The next day, May 21, notwithstanding the guides' promises that they would reach the Red River before nightfall, they found plenty of embarrassments to impede their progress. They spent most of the day in fatiguing travel through wooded elevations dominated by "dwarfish post and black oaks." The forest had recently known fire, which meant less underbrush to negotiate but also resulted in scorched tree limbs that were surprisingly effective barriers to travel.

The next morning, however, the guides discovered a familiar path that led them to the vicinity where a dozen or so families had come to form new lives. Although Major Bradford informed the squatters that they must depart this land reserved for the Indians, families such as the Styleses were hospitable and offered the travelers, tired of venison and jerky, fresh "milk and butter." William Styles's farm became their center of operations. Styles told them he had recently arrived here after taking a similar rocky, forested path through the Kiamichi wilderness—but he had a family, a wagon, and a mother-in-law who was "blind, and 90 years of age!" Styles lived near the Kiamichi, about a half dozen miles from the Red. The Red valley was a well-watered prairie, ablaze in the glory of spring. "Nothing could at this season," Nuttall exclaimed, "exceed the beauty of these plains, enamelled with such an uncommon variety

Kiamichi Wilderness, Spring 1819

of flowers of vivid tints, possessing all the brilliancy of tropical productions."

The color of the Red befit its name. The water was filled with the runoff of spring rains; it was "turgid," broad, hardly potable. The Red was a river with a long history. The Spanish had for centuries fought for control of the Red. They had without much success tried to form alliances with the native Comanche, Apache, Caddo, and Wichita tribes against the growing influence of first the French, then the English, finally the Americans. Spain was reluctant to recognize the Louisiana Purchase and aggressively defended its right to control the Red River valley. The Adams-Onis Treaty during this year of 1819 laid to rest the contentious issue of the boundary between the United States and Spanish territory. Although the Red would now be the official boundary, it would not stop Americans such as James Long, eager to acquire control over the lands of Texas, from frequent, often violent incursions into Spanish (subsequently Mexican) holdings.

Nuttall was, however, much more interested in natural rather than human history. On his peregrinations about the Red valley Nuttall discovered a new species of elm, the cedar elm (*Ulmus crassifolia*), a name he apparently chose because it grew in areas populated with the rock cedar (*Juniperus ashei*), a small juniper tree. He found the bow wood, or Osage orange (*Maclura pomifera*), named for his friend William Maclure, the Philadelphia geologist. Osage orange has large, lovely looking green fruits that are, unfortunately, inedible. He also observed the eastern coral snake (*Micrurus fulvius*), a beautiful serpent of interspersed red, black, and yellow bands. Also poisonous was the water moccasin (*Agkistrodon piscivorus*): "it is nearly black, two or three feet long, and thick in proportion, the head triangular and compressed at the sides."

The troops, guides, and naturalist paralleled the Red upriver, searching for squatters to warn out. They passed through the "Horse-prairie, 15 miles above the mouth of the Kiamesha," at its confluence with the Red. They spied a few mavericks that gave the prairie its name. Nuttall learned that wild boars (*Sus*) roamed

The Land between the Rivers

the valley. He was so overwhelmed by the variety of plants he became obsessed with botanizing, so much so that he separated from the others, became lost, and ended up seven miles from the Styles house, the agreed-upon rendezvous. The soldiers meanwhile returned to the Styleses' to prepare for the return journey to Fort Smith. As dusk settled Nuttall found an accommodating host, Mr. Davis, who informed the absentminded scientist of his location, on Gates Creek, several miles from the Kiamichi River and the Styleses'.

Having left his horse to graze during the night, Nuttall could not find the animal upon rising the next morning. By the time he found the horse (several miles away) and took the path to the Styles house, he discovered that the soldiers had recently departed. Styles's son led Nuttall in pursuit "for about 10 miles through a horrid brake of scrubby oaks, but all to no purpose." Stranded, Nuttall returned to the Styleses', botanizing as he went! "My botanical acquisitions in the prairies," he noted in his journal, "proved . . . so interesting as almost to make me forget my situation, cast away as I was amidst the refuse of society, without money and without acquaintance."

Thomas Nuttall reluctantly enjoyed the hospitality of William Styles and family for a fortnight. He spent the days wandering the prairies adjacent to the Styleses' land, awaiting the prospect of a hunting party or such going toward the Arkansas valley and Fort Smith. The prairie grass was "knee high," and among it were "flowers of the most splendid hues," which gave "the appearance of a magnificent garden." Nuttall, "scarcely" able to describe what he saw, resorted to science. Excitedly he discovered new species of already known genera of plants. The prairies "were perfectly gilded with millions of the flowers of *Rudbeckia amplexicaulis*," a new species of the coneflower. From another visual perspective the prairies looked snow covered, such were the numbers of white-flowered coriander (*Coriandrum*). Also prevalent in great numbers was a new species of the genus centaury (*Centarium americana*), the European species long used in herbal medicine. The days were hot, and rain was scarce: "All the lesser brooks and

Kiamichi Wilderness, Spring 1819

neighbouring springs were now already dried up, and the arid places appeared quite scorched with the heat. Still there prevailed throughout these prairies, as over the sea, a refreshing breeze, which continued for the greatest part of the day." The heat and wind effectively drove away "musquitoes," a delight to any explorer.

June 6 was one of the highlights of Nuttall's Red River excursion: "I now, for the first time in my life, notwithstanding my long residence and peregrinations in North America, hearkened to the inimitable notes of the mocking-bird (*Turdus polyglottus*). After amusing itself in ludicrous imitations of other birds, perched on the topmost bough of a spreading elm, it at length broke forth into a strain of melody the most wild, varied, and pathetic, that ever I had heard from any thing less than human. In the midst of these enchanting strains, which gradually increased to loudness, it oftentimes flew upwards from the topmost twig, continuing its note as if overpowered by the sublimest ecstasy." The songs and behavior of the mockingbird henceforth became one of Nuttall's favorite studies. He wrote of it at length in his *Ornithology*, basing his description on communications from friends and correspondents, as well as his own observations made during his Arkansas journey and, years later, his journey to the southern states. Visiting Alabama in February 1830, Nuttall "heard the Mocking Bird" imitate in short order "the Carolina Woodpecker," "Carolina Wren," "Cardinal bird," and the "Tufted Titmouse"—and in each case the imitation was more melodic than the original. Indeed, repeatedly Nuttall observed the mockingbird's call deceive individual birds into thinking that their mates were nigh. And yet, compared to the beauty and "brilliant plumage" of these others, the mockingbird is a dismal gray and white. The beauty of this "Orpheus" is its song, which it performs with glee and mockery, chattering and scolding, dancing in the air, swooping about in fun or intimidation. Nuttall grew so familiar with this bird that he kept one caged to observe, even communicate with it. It was a delightful friend, absorbing the attention of the observer. Nuttall hardly needed the cage, as he admitted in his *Ornithology*.

The Land between the Rivers

The mockingbird often befriends a human and his habitation; it takes a while, as the bird seemingly ponders the character of the host and his cottage. Soon he feels sufficiently comfortable to approach closer, to chatter nearby and entertain. The mockingbird builds his nest in a bush, preferably next to the cottage. Nuttall believed that the bird yearns for the closeness of another home to his own. Indeed, Nuttall thought he could read the feelings of the bird, come to know it, even empathize with something so wild and free. One detects in his descriptions of the mockingbird a yearning for the frontier and the indescribable quality of the wilderness that simultaneously attracts and repels, pleases and disappoints, nurtures yet destroys. Ironically, one finds the same creative and destructive phenomenon imposed upon freedom by civilization. As Nuttall explained in his *Ornithology:* "It is impossible to listen to these Orphean strains, when delivered by a superior songster in his native woods, without being deeply affected, and almost riveted to the spot, by the complicated feelings of wonder and delight, in which, from the graceful and sympathetic action, as well as enchanting voice of the performer, the eye is no less gratified than the ear. It is, however, painful to reflect, that these extraordinary powers of nature, exercised with so much generous freedom in a state of confinement, are not calculated for long endurance, and after this most wonderful and interesting prisoner has survived for 6 or 7 years, blindness often terminates his gay career; and thus shut out from the cheering light, the solace of his lonely but active exsitence [*sic*], he now, after a time, droops in silent sadness and dies."

Nuttall, having learned of an intended journey to Fort Smith by a party from the squatter community along the Red River, visited the settlers and arranged to join them. His initial impression of the squatters and their sort was that they had a questionable "moral character," there "being many of them renegadoes from justice, and such as have forfeited the esteem of civilized society." But once the three men arrived as promised on Sunday ("a day generally chosen by these hunters and voyageurs on which to commence their jour-

Kiamichi Wilderness, Spring 1819

neys"), Nuttall found them to be, like William Styles, "men of diligence and industry." They camped the evening of June 14 next to "a brook, beneath the shade of the forest, and under the serene canopy of a cloudless sky." The squatters, having just been warned away from the Red River valley by Major Bradford to make room for proposed Indian settlement, were naturally upset with the U.S. government for preferring the Indians over themselves. These three men were intent on "the recovery of horses stolen from them by the Cherokees," who had long hunted in this land between rivers and who now resided on former Osage lands in Arkansas Territory.

Notwithstanding directions from Styles, as well as a good initial buffalo trace, the second day out they lost the trace and their bearings and spent a considerable amount of time struggling through utter wilderness. "We passed and repassed several terrific ridges," Nuttall wrote, "over which our horses could scarcely keep their feet, and which were, besides, so overgrown with bushes and trees half-burnt, with ragged limbs, that every thing about us, not of leather, was lashed and torn to pieces." They kept to the Kiamichi River as providing the best bearing for travel, generally following it upriver toward and through the mountains. Nuttall, of course, continued to botanize, discovering at a lake near the Kiamichi plants that he heretofore had thought were exclusively coastal, such as the yellow pond lily (*Nuphar advena*), as well as the pickerelweed (*Pontederia cordata*), an edible plant that went perfectly well with its namesake, the smooth-skinned, long and sleek pickerel (*Esox americanus vermiculatus*). A profusion of wildflowers appearing after a fatiguing journey "through a succession of horrid, labyrinthine thickets and cane-brakes" served as a delightful and refreshing cordial for the tired men. On June 18 the explorers journeyed toward a ridge that formed the watershed separating the Arkansas and Red river valleys.

As the heat of summer approached so did, with increasing frequency, the variety of flies that irritate man and beast. Exceedingly large horseflies—cleg flies (*Haemotopota*)—so plagued the horses at rest and grazing during a noon break that they stampeded in search

109

of a remedy, finding it about five miles away in the waters of the Kiamichi, where the suffering animals were finally discovered after a tiresome search by Nuttall and his companions.

Over the course of this journey the men repeatedly realized the want of a guide. Having thought they knew where a notch lay in the mountains, they nevertheless missed it, finding themselves at the highest elevation, "high as any part of the Blue Ridge, through thickets of dwarf oaks (*Quercus chinquapin, Q. montana, and Q. alba*), none of them scarcely exceeding the height of a man." The toil of travel was unremitting, until at length they stumbled upon "the wagon trace, so recently trod by the major's party." Soon they entered the Poteau River valley, following the trail along prairies between low, rocky ridges: "The prairies were now horribly infested with cleg flies, which tormented and stimulated our horses into a perpetual gallop." On June 21, along with the summer solstice, they arrived at Belle Point.

14 The Verdigris, July 1819

Thomas Nuttall waited about a fortnight for the opportunity to ascend the Arkansas River to Three Forks. He spent his time botanizing in surrounding prairies; ordering his growing collection of seeds and specimens; and conversing with Dr. Russell and local hunters about the forbidding territory to the west, the character of the Osages, and the best routes by which to approach the Rocky Mountains. At length Nuttall persuaded Joseph Bogy, the Arkansas trader, to allow him space in the pirogue that was to ascend the Arkansas to Bogy's trading post situated on the Verdigris River. Nuttall settled into the hollowed-out log fitted with oar-locks, rowed by the French voyageurs.

The river was wide and winding, enclosed by sandy beaches and, at times, bluffs and cliffs. Great islands of sand, many with lush vegetation, forced the river into varying forks and occasionally formed obstacles to challenge the boatmen's skill. The journeyers made the sandbars their ports along the Arkansas. Here they could rest, relatively secure from the mosquitoes and ticks of the forest. The sandbars even hosted new and unusual flora, such as the rose moss (*Portulaca pilosa*), its bright pink flowers suggesting a habitat radically different from the hot dry sand. The sand also hosted an environment thought sufficiently secure by the painted turtle (*Chrysemys picta*) to deposit her eggs deep in the sand. The voyageurs, cognizant of her habits, "amused themselves," Nuttall wrote, "searching for turtle's eggs, which the females deposit in the sand at the depth of about eight or ten inches, and then abandon their hatching to the genial heat of the sun. They are spherical, covered with flexible skin, and considered wholesome food." Also tasty were sand plums (*Prunus angustifolia*), raspberries (*Rubus occidentalis*), and hazelnuts.

111

The Land between the Rivers

On July 11 the voyageurs and scientist passed the mouth of the Canadian, a significant river feeding the Arkansas. Nuttall had accumulated much, often erroneous information about this river: "Its main south branch sources with Red river, while another considerable body keeps a western course through the saline plains, where it becomes partially absorbed in the sands of the desert, but afterwards continues towards Santa Fe or the Del Norte." The source of the Canadian is in the mountains of eastern New Mexico, while the Red begins much further east, in the plains of the Texas Panhandle. No fork of the Canadian proceeds west toward Santa Fe. Although Zebulon Pike had discovered the source of the Arkansas River in southern Colorado, the sources of the Canadian and Red were still unknown. Nuttall's error, that the Red and Canadian have a similar source, had several origins: the tales of hunters, speculation of soldiers, and the similarity in color and content of the Canadian and Red Rivers. Both possess the deep red color and high saline content of rivers imbued with the runoff of the red clay soil of the west. The Arkansas, however, is clearer, in part because its next few tributaries are rivers flowing from the more forested foothills of the Ozark Mountains. The first such river the travelers came to was the Illinois, "a considerable stream of clear water." A few miles up the Arkansas from the mouth of the Illinois is Webber's Falls, "a cascade of two or three feet perpendicular fall." In attempting to ascend the fall the pirogue grounded; the boatmen decided to back off, camp on the shore, and see whether or not the river might rise overnight. It did not, but the wind changed, which filled the single sail of the pirogue sufficiently to help the voyageurs pole and row it past the white water.

The unusual southeast wind kept up for several days, which allowed their rapid progress up the Arkansas toward Three Forks. The region had known little rain for over a month; the small creeks inland were drying out, which forced forest animals to the river. Such trees that grew on the banks of the Arkansas were dwarfed by the increasingly dry conditions. The cottonwood, less imposing than along rivers further east, was plentiful, as was the American

The Verdigris, July 1819

elm (*Ulmus americana*), box elder (*Acer negundo*), its cousin the curled maple (*A. dasycarpon*), and different species of ash (*Fraxinus*). The Arkansas was in places "deep, . . . clear, green, and limpid," which indicated that they were near Three Forks and the Grand and Verdigris Rivers. The latter river supplied the greenish tint to the water. On July 14 they passed by the mouth of the Grand (Neosho) River and shortly thereafter the Verdigris, up which they went.

The Three Forks, for centuries a crossroads of trade for Indians, French, Spanish, and Americans, hosted several significant trading posts, which attracted the few squatters of the area, hunters and trappers, and the Osage Indians. Joseph Bogy's trading post was on the west bank of the Verdigris just above its mouth; Captain Nathaniel Pryor's trading post was next door. Directly across the river was the post of Henry Barbour and George Brand. Just below Barbour and Brand's was Hugh Glenn's post. Hugh Glenn was the communicative trader that Nuttall had met eight months earlier at Cincinnati. Glenn was absent from his post, but Captain Pryor was friendly and accommodating to the newcomer Nuttall. Pryor acted as tour guide for Nuttall on the morning of July 14, taking the scientist into the fertile U-shaped peninsula between the Verdigris River and the Grand River. The locals called the peninsula French Point. Well-used Indian trails crisscrossed French Point. The remnants of Indian villages were ubiquitous. The rich alluvial soil hosted dense cane near the rivers and unexpectedly large trees, such as the scarlet oak (*Quercus coccinea*), typically found in forests more to the east; white ash (*Fraxinus americana*), an important part of the Osage materia medica; and hackberry (*Celtis occidentalis*), a tree often found near the nettle bush (*Urtica*), the leaves of which, steeped in hot water, form a tea that Indians and settlers alike used for ailments ranging from arthritis to intestinal and urinary problems. According to Nuttall, hunters made rope from nettle roots.

The two men, Pryor and Nuttall, wandered north, away from the rivers, into "the great Osage prairie, more than 60 miles in length." The botanist here observed the leadplant (*Amorpha canescens*), a tall,

The Land between the Rivers

hardy, pretty prairie plant used by the Indians for tea and as an ingredient for the calumet. Nuttall also discovered "a new species of *Helianthus*," the sunflower, an important source of food, oil, and medicine for Indians and settlers alike, besides being an astonishingly beautiful wildflower. The sunflower thrives in the heat of July and August on the prairie.

Pryor led Nuttall west to the falls of the Verdigris, where according to some traditions De Soto had made his camp of Autiamque in 1541. There were three sets of falls, the first being about five miles above the river's mouth. The river hosted buffalofish (*Megastomatobus cyprinella*), as well as the long-nose gar (*Lepisosteus osseus*). Along the hot summer riverbanks there was a profusion of insects and spiders; reptiles such as the five-lined skink (*Eumeces fasciatus*) preyed upon them.

The Arkansas Territory was, like most frontier regions, heavily dependent upon salt springs. Nuttall journeyed up the Grand River by dugout canoe with two young guides to see a local saltworks. The Grand was shallow yet wonderfully clear, the water cool, like most rivers fed by nearby hills. Rapids caused by low water running over gravelly bars obstructed most boats. Along the way Nuttall discovered the American smoke tree (*Cotinus obovatus*) and, in the surrounding forests, several black bears. The bears lived among trees that provided much wholesome food, such as the Kentucky coffee tree (*Gymnocladus dioicus*), the swamp post oak (*Quercus lyrata*), and the pecan (*Carya illinoensis*), as well as various hickories, elms, birches, and ashes. The saltworks had until recently been very productive, producing "120 bushels of salt" weekly—worth enough money to encourage theft and murder. This had been the case several months earlier, when one owner of the works killed his partner. His two accomplices included a hunter, William Childers, whom Nuttall had seen the previous March at the Cadron settlement on the Arkansas and had tried to hire as a guide, not knowing he was a murderer on the run.

On July 20 Nuttall set out on his own overland across the prairie to Three Forks. Having felt the ague approaching, he took a large

114

dose of boneset (*Eupatorium perfoliatum*), which had a laxative effect and helped him to feel strong enough to pursue the journey. The prairie grass was waist deep, mixed with briars and covered with honeydew, which "was so universally abundant, that my mockasins and pantaloons were soaked as with oil." The entire day Nuttall labored through this growth, "wrapt in primeval solitude," toward his goal for the day, a small hill in the distance, in the direction of travel "south by west." Nuttall ascended the hill to see "from its summit, the wide and verdant plain." After this he journeyed until dark, when without "fire, food, or water," he lay down in the grass to sleep: "The crickets, grashoppers [*sic*], catidids, and stocking weavers, as they are familiarly called, kept up such a loud and shrill crepitation, as to prove extremely irksome, and almost stunning to the ears." Nevertheless, Nuttall fell asleep and "slept . . . in comfort, and was scarcely at all molested by musquetoes."

Having returned to the trading posts on the Verdigris, Nuttall was present during the last week of July, when the Osages returned from a summer buffalo hunt up the Arkansas River. Nuttall was genuinely interested in American Indian tribes and lacked the common assumption found among white Americans that Indians were savages. However, during his many journeys some tribes impressed him more than others; some seemed more "civilized" than others. Nuttall had mixed impressions of the Osages. Stories among whites of the Arkansas Territory related their cruelty and savagery. Nuttall had not seen any of this. He did find them to be heavily dependent upon white traders for metals, tobacco, and whiskey. Such dependence appeared an utter contradiction to what Nuttall assumed should be the Indian's inherent ability to find all that he needed to make his life pleasant among the vast multiplicity of nature. Yet their strength and stamina astonished Nuttall, who thought a twenty-mile hike was quite an accomplishment until he heard of the Osages' sixty-mile hikes in one day. Nuttall learned "from the Osage interpreter, of whom I made the inquiry, . . . that, in common with many other Indians, as might be supposed from their wandering habits and exposure to the elements, they are not unacquainted with some peculiar characters and configurations of the stars. Habitual observation had taught them that the pole star remains stationary, and that all the others appear to revolve around it; they were acquainted with the Pleiades, for which they had a peculiar name, and remarked the three stars of Orion's belt. The planet Venus they recognised as the Lucifier [*sic*] or harbinger of day; and, as well as the Europeans, they called the Galaxy the heavenly path or celestial road. The filling and waning of the moon regulated their minor periods of time, and the number of moons, accompanied by the concomitant phenomena of the seasons, pointed out the natural duration of the year." The Osages were a religious people, worshipping the Great Spirit, lamenting the dead, and practicing "rigid fasts, . . . a kind of penance, by which they disciplined themselves for disasters, and supplicated the pity and favour of heaven." They were affectionate toward their spouses, though men practiced polygamy; indeed, women had few rights and did most of the work. The Osages despised dishonesty and dissimulation and thought that achievements in war or the hunt were an individual's greatest calling. And yet they had a reputation of being irreformable thieves.

Nuttall discovered Osage thievery firsthand on several occasions. About the first of August he experienced a return of the ague; while he was ill, an Osage "contrived to rob me of the only penknife in my possession, and my pocket microscope." The Osage chief, Talai, made a speech before his people condemning such practices, but the items never turned up. A few days later another Osage offered to trade his horse for Nuttall's recently purchased mare. The Indian horse "was not worth possession," so Nuttall refused. Nuttall's "mare was at this time feeding across the Verdigris." The Indian departed, but Nuttall suspected his intentions and so with another man crossed the Verdigris just in time to stop the Indian from stealing his horse. The Indian in response "began to speak in a submissive tone," but so intimidated were the traders and trappers by the Osages and their authority that Nuttall found himself obligated "to bestow a present upon the villain."

Thomas Nuttall would have more to do with the Osages as the

The Verdigris, July 1819

summer grew hotter, water became sparse on the prairies, food was more difficult to find, and competition for the limited resources of the land between the rivers bred conflict between newcomers and apparent trespassers and local Indian tribes. Nuttall soon would begin a new journey overland that would last five weeks, bring him near to death, and involve further interactions with the Osages of the Arkansas River.

15 Little North Fork, August 1819

Lee the hunter and trapper was also at Three Forks the summer of 1819. Precisely when he and Nuttall first met is unclear. Perhaps Lee was the unnamed hunter in Nuttall's journal whom the scientist "accompanied . . . about 9 or 10 miles over the alluvial lands of Grand river" on July 28. In his journal entry for August 6 Nuttall noted that he "had purchased of Mr. Lee" a horse in preparation for the land journey that began five days later. Clearly, hunter and scientist had gained some familiarity with each other the several weeks before the beginning (on August 11) of their journey west across the prairies. Nuttall sized up Lee as an experienced hunter who knew the western prairies and river valleys that must be crossed and ascended to reach the Rocky Mountains. Lee was one who could live off the land, locate sources of water, obtain meat, and know what in the vegetable kingdom one could eat when game was scarce. Lee claimed familiarity with the Osages, Cherokees, Comanches, and Apaches who lived along the Red, Canadian, and Arkansas Rivers. He claimed as well to have ascended some of these rivers as far west as the foothills of the Rockies.

Lee, meanwhile, was astonished to find the likes of Thomas Nuttall in such a hard country. In some respects Nuttall appeared as a greenhorn, completely out of place in the wilderness. Observing the scientist, having even brief interaction with him, Lee learned of the Englishman's unusual preoccupation with plants, which he would examine closely, taste, and carefully preserve. The dominant concerns of those who frequented Three Forks were horses, beaver, the Indians, river levels, and the comings and goings of trappers, hunter, traders, voyageurs, Indians, soldiers, and desperadoes. Here was a strange one indeed, who was clearly not a hunter, much less a

Little North Fork, August 1819

marksman; who was articulate, not coarse in his speech; who refrained from liquor, tobacco, and other vices; who appeared small and delicate, hardly able to withstand the toil, suffering, and danger of the West; and who was suffering from symptoms of the ague, as were many at Three Forks. And yet here he was, asking whether or not he could join the hunter on his ensuing journey across the plains, offering to pay him for his trouble.

Lee assented, and we might ask why. Trapping was a solitary, lonely business, which Lee had pursued for eight years. During this time he had seen the contrary nature of this land between rivers— the extremes in the weather; the variation in game depending on the season and the climate, in particular the amount of rainfall in a given year; and the extremes in the behavior of the Indians who inhabited and hunted the valleys of the Red, Canadian, and Arkansas. Hunters such as Lee at times got along quite well with the Native Americans. They were, of course, similar in disposition, lifestyle, and employment. This was, ironically, more an issue of concern for the Indians than the white hunters. The hunter sought the same pelts, water, and food as the Indians. Lee, for example, discovered the consequences of successful trapping the summer of 1818, when he had the unfortunate experience of meeting a hunting party of Cherokees along the Canadian River. He received the same treatment dished out to most such hunters: he was stripped of his horse, weapons, furs, supplies, and clothes. Lee was sufficiently experienced to accept the gauntlet and not to resist. Indians tended to respect courageous endurance of suffering rather than the weakness of desperation. At the same time, on the Arkansas, the Osages punished another hunter, James Macfarlane, who was found with the enemy Pawnees, but in quite a different way: "They seized upon him, put out his eyes, and then goaded him along for several miles with sharpened canes, thus protracting his death by torture, until one of them, through compassion, put an end to his existence by the tomahawk."

Into this territory of warring tribes, jealousy, murder, and lawlessness—a human society as seemingly chaotic as the natural envi-

The Land between the Rivers

ronment of unrelenting heat, random thunderstorms, dust, and hoards of flies and other vermin—Thomas Nuttall, the English scientist, member of the American Philosophical Society, associate of the Academy of Natural Sciences, member of the Linnaean Society of London, author of one of the best treatments of America's flora available, as well as one of the leading minds of the Philadelphia intellectual community, set forth on August 11, 1819, accompanied by a trapper who, we may suppose, was ignorant of and apathetic about Nuttall's scientific and botanical concerns, who was a simple frontiersman, perhaps illiterate, with limited knowledge of the world beyond the land between the rivers. Lee did, however, purport to know the highlands to the west, near the sources of the Red, Canadian, and Arkansas Rivers. Nuttall, at Arkansas Post six months earlier, had been persuaded by Joseph Bogy that the Canadian was the best river to ascend to the mountains to find the source of the Red River, which was still unknown to Americans. But at Three Forks Nuttall changed his mind, perhaps on the advice of Lee, who suggested that they depart from the Arkansas and set out overland to the north forks of the Canadian River, which would take them near to the Cimarron River, one of the tributaries of the Arkansas. Nuttall could ascend the Cimarron to achieve his goal of botanizing in the Rocky Mountains.

They set out, then, in a west-southwesterly direction across the prairies of what would in time be central Oklahoma. The cottonwoods and other trees to the north signaled the direction of the Arkansas as it went on its northwesterly path. Lee pointed out Osage traces where, he guessed, "2 or 300 men and their families" traveled to and from the Arkansas to hunt. Lee convinced Nuttall that they should take care, for "to meet the Indians" would mean that "they would, probably, rob us of our horses, if not of our baggage, and illtreat us besides, according to the dictates of their caprice and the object of their party." The heat was oppressive, the horseflies everywhere, the wind hot from the south, and fresh water scarce. Nuttall, still battling the effects of the ague, was driven by thirst the second day out to drink from a stagnant creek of brown, warm water. He immediately became ill and experienced repeated vomiting, such "that it was with difficulty I kept upon my horse."

After camping next to a creek, where Lee set his beaver traps, the next morning, July 13, Nuttall's health returned, if temporarily. The day's journey took them across grassy plains destitute of wildlife if not insects. They crossed two shallow creeks before arriving at "a considerable rivulet of clear and still water, deep enough to swim our horses." The river helped to cool an otherwise hot day and provide diversion for the eyes, which Nuttall complained ached from "the dazzling light of the prairies." Lee's makeshift trail passed by more Osage traces, Indian burial mounds, "the burrow of a badger" (*Taxidea taxus*), and the small mounds and open holes of hornets' nests (those of the yellow jacket, *Vespula,* and the cicada killer, *Sphecius speciosus*). But all was quiet and still, save the blowing wind, the rustling grass, and the tormenting flies.

They camped and rested the next day on the banks of the Little North (Deep) Fork of the Canadian River. Lee and Nuttall were about forty miles from its confluence with the North Fork of the Canadian, which was itself a tributary of the Canadian River, the confluence being about fifty miles further east. Lee's horse, "being from the first totally unfit to travel from a large wound on its back," required rest. Likewise Nuttall had become terribly ill with a high fever made worse by the afternoon sun. He was weak and delirious; it was all he could do to "crawl into the shade" and vomit. Lee, undeterred, set his traps, catching four beavers (*Castor canadensis*) the first night. He informed Nuttall that "scarcely any thing is now employed for bait but the musk or castoreum of the animal itself. As they live in community, they are jealous and hostile to strangers of their own species, and following the scent of the bait, are deceived into the trap."

The banks of the Little North Fork now became their home as Nuttall's fever continued unabated, accompanied by the chills of ague and diarrhea. Their food began to dwindle because of the activity of the "green blow-flies, attracted by the meat brought to our camp," which "exceeded every thing that can be conceived. They

The Land between the Rivers

filled even our clothes with maggots, and penetrated the wounds of our horses, so as to render them almost incurable." After several days they moved five miles downstream "for the purpose of trapping." Nuttall's illness was becoming more desperate. He could hardly sit a horse and was in and out of delirium. Lee had the sense to realize their dilemma and "suggested the propriety of our returning to the Verdigris, before I became so weak as to render it impossible; but the idea of returning filled me with deep regret, and I felt strongly opposed to it whatever might be the consequences."

The next week on the prairie featured much of the same. Nuttall's growing weakness, as well as Lee's frustration at their inactivity and problems with his horse, resulted in a growing rift between the two men. Nuttall began to doubt Lee's ability to guide them, while Lee was angered over the scientist's obsession with continuing the journey, even as he was nearing death. The increasing signs of an Indian presence frightened Lee, who was more aware than Nuttall of what such a confrontation could mean. Food was growing short. Lee, however, discovered a bee tree along the river banks, which provided the only food Nuttall could keep down. More fortunate was the cool front that roared through the region the night of August 19. The storm brought with it fresh water and cooler weather. The skies remained cloudy for several days, which gave Nuttall a chance to recover.

On August 24 Nuttall felt sufficiently like himself to botanize, if briefly. Moving slowly west, they "passed by three or four enormous ponds grown up with aquatics, among which were thousands of acres of the great pond lily (*Cyamus luteus*), amidst which grew also the *Thalia dealbata*, now in flower, and, for the first time, I saw the *Zizania miliacea* of Michaux." The former, Nuttall's pond lily, was the water chinaquin, or American lotus (*Nelumbo lutea*), its large, beautiful yellow flower buzzing with bees. The *Zizania* was wild rice, a favorite repast of the red-winged blackbird (*Agelaius phoeniceus*). Nuttall was disappointed to find, however, that neither blackbirds nor any other birds were willing to brave the heat of the prairie.

Little North Fork, August 1819

On the pond great banks of ragweed (*Ambrosia trifada*) "were higher than my head on horseback." Whether or not Nuttall knew that ragweed tea could help alleviate fever, or that Indian children chewed the root to reduce anxiety, is not clear. But Nuttall needed help on both scores.

The final days of August brought to the scientist, helpless in this harsh environment, feelings of despondency and attacks of anxiety. "Gloom" filled his days, "misery and delirium" his nights. On August 30 "my mind became so unaccountably affected with horror and distraction, that, for a time, it was impossible to proceed." Frozen with fear, unable to move, Nuttall plopped down in the prairie; what Lee thought we can only imagine. Nuttall could move neither forward nor back in place or in time. When each moment heralds disaster, why move, why live? Having begun the day departing the freshwater and ponds of the Little North Fork valley and proceeding over the dry prairies, Nuttall became overwhelmed with the solitude, the emptiness, the loneliness of the open skies and distant horizon. There seemed to be no escape from the sameness of the prairies. The Rocky Mountains were an eternity away. Philadelphia, England, civilization, were likewise at an untold distance. Despair convinced Nuttall of his impending death far away from familiar sights and people. The hard beauty of the land prickled his mind; the landscape scarred his being. The sun scorched, and then the rain pelted; the unquenchable heat gave way to unbearable cold; the wonder and beauty of grasses, flowers, and trees masked the destruction hidden therein. What is it that allows death to overwhelm life? More amazing, how does life withstand the temptations of death? What *will* within Nuttall tipped the scales away from death toward life?

September brought with it hope. Nuttall, recovering the urge to proceed toward journey's end, entertaining the vain hope that he would indeed reach the Rockies, castigated Lee for not knowing where they were. Lee, using his wits rather than maps, thought that the Cimarron (Salt) River was nearby. Lee was, however, "entirely

The Land between the Rivers

deceived"—at least that was the belief of Nuttall, who branded Lee in the pages of his journal with the charge of being "ignorant of the country."

Thomas Nuttall rather than Mr. Lee was deceived. Lee was ignorant according to Nuttall's standards of latitude and longitude, precise distances in miles or rods, degrees on the compass, plots on a map. Lee was not ignorant, however, of the lay of the land, the feel of the wind, the hidden signs of water, the otherwise unseen river valley—just up ahead. Lee intuited his route as he did all of life's journeys. They would get there; when or how was less certain than his basic feeling that the Cimarron was nearby and that he would stumble upon it—if not today, then tomorrow or perhaps next week. Lee's itinerary of nature was quite different from Nuttall's itinerary of science.

The first week of September 1819 a doubtful Nuttall and a hopeful Lee continued their journey north, through a slightly hilly country of frequent streams of cool, potable water and "an endless scrubby forest of dwarfish oaks." The cross-timbers of central Oklahoma feature trees such as the chinkapin oak (*Quercus muehlenbergii*), post oak (*Q. stellata*), black oak (*Q. velutina*), and blackjack oak (*Q. marilandica*). Lee, either to help the ailing scientist or because of an incipient interest in botany, identified and picked a specimen of rose moss (*Portulaca pilosa*). Lee continued to trap for beaver with mixed success. One creek they came to was dry because of the efforts of an industrious beaver, which had dammed the creek to form a freshwater pond, the banks of which hosted tall grass and small willows and cottonwoods. Nuttall, still greatly fatigued and not altogether well, yet better, pushed himself to make "26 miles through the same kind of deeply undulated country." Eventually, near the end of the day, Lee spied a sign that he was not so far off track as Nuttall might have assumed. The northern horizon had a hazy appearance, which Lee declared was blowing sand—a sign of the banks of a river. By dusk they had reached the Salt River, or the "First Red Fork"; today it is known as the Cimarron.

16 The Cimarron, September 1819

The trapper and hunter Mr. Lee reached the Cimarron River wondering what had compelled him to join forces with the relentless botanizer Thomas Nuttall. Even sickness to the point of death could not prevent Nuttall from pursuing his fading goal of reaching the Rockies. Lee had relented to botanize for his ill companion, had fed him, and in a rough and tumble way had nurtured the scientist back to health sufficient for survival. At least Nuttall could stay on his horse for a day's ride, which had not been the case the preceding week.

Nuttall, however, reached the Cimarron entertaining feelings of new hope: "Its first view appeared beautifully contrasted with the broken and sterile country through which we had been travelling. The banks of cotton-wood (*Populus monilifera*), bordered by the even beach, resembled a verdant garden in panorama view." After a week on the Cimarron Nuttall's opinion would change.

The men's horses also appeared thankful, as they lapped the river water that Lee and Nuttall found "nauseous and impotably saline." Fresh water was not to be seen, which added to Lee's portrayal of the Cimarron; upstream, he said, were great sand dunes. Nuttall realized that they had come not so much to a garden as to an oasis in a land "of sterility little short of the African deserts." Such had been the christening that Zebulon Pike had given this land between rivers in his published journals. Subsequent explorers, too, such as Stephen Long, would agree with the image of this land as a desert, which was inaccurate though understandable based on the experiences of men trying to live in a place that is stingy and frequently forbidding.

Still entertaining an idea of pursuing the trek west along the

The Land between the Rivers

Cimarron to the high country—an absurd notion based on the tenuous health of Thomas Nuttall and Mr. Lee's horse—the two sojourners spent September 4 journeying up the Cimarron, "crossing it from point to point." The Cimarron, which rises in eastern New Mexico, is red and sluggish. Great expanses of sand, carved into irregular waves by the wind and river currents, mark the river's course. Signs of the Osages, the heat of the sandy beach, the lack of fresh water, and hardly any food, as well as illness, convinced the two to return, to journey down the Cimarron to the Arkansas and back to Three Forks. Even this plan had a kink in it, as Lee's horse, clearly destined from the start never to finish the journey, "got into a mirey gully, and could not be extricated." Having seen that the Cimarron would not be an accommodating host, with its heat, salt, and quicksand, the frontiersman decided to halt, rest, and fashion a dugout canoe from the thickest cottonwood trunk he could find. This was not an easy undertaking. Although cottonwoods were the largest trees growing along the banks of the Cimarron, they were still small in stature because of the sandy soil and dry climate. One tree, however, was found to be sufficiently thick. Its rough bark gave way quickly to Lee's axe, though its hard wood was not so accommodating. Once the tree was felled, Lee alternately chopped, burned, and scraped the wood to form a hollow arc in which to stow his gear, his pelts, and himself. After he had fashioned a paddle out of the same wood, Lee's creation was just large enough and "so exactly answered his purpose that it would have sunk with any additional loading."

Lee navigated the gentle current while keeping in sight his companion, Nuttall, who paralleled the river on horseback. What Lee experienced firsthand, Nuttall experienced vicariously. Amid the hot sand of the shore he yearned for his own canoe, which if it was not a steadier at least it was a quicker conveyance than a horse without a trail. But being on horseback allowed Nuttall the chance, at least, to reflect upon the river, to gaze upon its beauty, to imagine its terror. Summer drought meant shallow water more salty than during spring rains. Nuttall, forced at times to drink from the

The Cimarron, September 1819

Cimarron ("which always proved cathartic"), could not have agreed with James Wilkinson's assessment twelve years earlier that the river was potable.

In spite of little water, only beaver's tail for food, intense heat, and lingering illness, Nuttall nevertheless took to botanizing. He discovered a species of honeysuckle (*Lonicera*); a species of gentian (*Gentiana*), a useful medicine in treating the ague; and a species of sea oats (*Uniola*), typically found on the seashore, far away from the rivers of the inland prairies.

Midday on September 9, after one and a half days on the river, they reached the mouth at its confluence with the Arkansas River. The Arkansas was broad and shallow, like the Cimarron, yet its water was clearer, better to drink, if not altogether free from sediment. Nuttall looked about and saw the thick brush along the sandy shore, saw cottonwoods shading the uncertain path before him. He saw the alluvial lands rising from the water, the fertile soil deposited by a thousand floods. Even if all about him appeared marked with privation and suffering, the region "was not without beauty." Recalling the similar environs of the Ohio River, which Nuttall had descended almost a year earlier, he imagined such fertility might someday yield civilization from the wilderness. "I could not help indeed reflecting," he wrote in his journal, "on the inhospitality of this pathless desert, which will one day perhaps give way to the blessings of civilization." Such was the uncertain prediction of the future city of Tulsa by the weary traveler.

Along both the Cimarron and the Arkansas were frequent signs of the Osage Indians. Game was scarce, there were fresh tracks of people and animals, and smoke sometimes billowed from behind distant bluffs. Lee thought that the Osages were clearly on the move, engaged in their "summer hunt." The Osages simultaneously terrified, mystified, and fascinated Nuttall. Perhaps during the many days riding on the sandy banks of the Cimarron Nuttall reflected on the Osages, on their dignity and courage, their chiefs of a modest and urbane turn, the people pious in religious devotion and loyal to family and spouses. On the other hand he had discovered them to be at times untrustworthy, murderous, and vengeful, in part (he believed) because of their natural aboriginal character, in part because of their constant want of and hunger for the bare necessities of life, which drove them to thievery and warfare. If at Three Forks Nuttall had looked to the former, humane side of the Osages, now—descending the Arkansas alone on horseback with a companion in a nearby canoe—was hardly the time to expect courtesy from what appeared to be a natural enemy.

Nuttall's virgin trail often brought him an uncomfortable distance from the river and his companion. Thus, alone on September 10, forced to ride amid high bluffs that rose from the river, he spied an Indian stalking him, dodging "out of sight" when seen. "This wolfish behaviour," Nuttall later recalled, "it may be certain, was not calculated to give me any very favourable anticipation of our reception." Nuttall and Lee nevertheless maintained the illusion that it was possible to avoid a direct confrontation. The next day, however, necessity forced itself upon their affairs: several Osage "families were encamped along the borders of the river." The water was sufficiently fresh to drink, though hunting produced nothing significant and the fishing was bad. The wayfarers intrigued the Osages, who "ran up to us with a confidence which was by no means reciprocal." Each side was doubtless suspicious of the other's intentions: Nuttall and Lee assumed the Osages would try to rob them— or worse; the Osages were spellbound to find one hard-bitten man naturally traveling by canoe and another delicate soul, clearly out of place, trying to navigate the pathless bluffs paralleling the river.

The leader of this small band of natives was old and blind and "not unknown to Mr. Lee," who had learned how best to befriend Indian chiefs; Lee fortunately had some tobacco in his possession, which he willingly shared with the chief, a gift "with which [the chief] appeared to be satisfied." The chief reciprocated by asking his guests to a dinner of "boiled maize, sweetened with the marmalade of pumpkins," which after a diet of beaver tail for the past week was a welcome repast for Nuttall and Lee. The chief "showed us a commendatory certificate which he had obtained at St. Louis."

The Cimarron, September 1819

Such pleasures did not last: "About the encampment there were a host of squaws, who were extremely impertinent. An old woman, resembling one of the imaginary witches [in] Macbeth, told me, with an air of insolence, that I must give her my horse for her daughter to ride on; I could walk;—that the Osages were numerous, and could soon take it from me." When the two wary travelers tried to depart, a dozen of the Indians, young and old, male and female—"even the blind chief"!—tried to steal what they could from Lee's canoe as well as the canoe itself. One young brave approached Nuttall and demanded his horse, "insisting that it was his." Nuttall, exasperated, "could no way satisfy his unfounded demand, but by giving him one of my blankets."

Finally having gotten away from the chief and his acquisitive people, the travelers found no relief, as two braves along the opposite bank of the Arkansas "hid themselves" at their approach, then "began to follow" the hunter and scientist "as secretly as possible." Nuttall recalled later that "they continued after us all the remainder of the day, till dark. We knew not whether they intended to kill or to rob us; and, endeavouring to elude their pursuit, we kept on in the night."

Help came from an unexpected quarter. The sky grew a menacing blue-green; windswept waves battered the shore; lightning blinded; thunder deafened. The two men knew the peril of the storm, but they welcomed its onset all the same. Lee paddled his dugout canoe furiously in apparent utter futility, considering that escape from the randomness and instantaneity of the lightning stroke was impossible—especially when the level plane of the river made a canoe and its occupant stand out, a perfect target. The situation of Nuttall was even worse. Nuttall urged his horse on in the storm, through sandy beaches, uncertain shallows, and the ever-present danger of quicksand. The unaccommodating riverbank forced man and horse to cross the river repeatedly. The scientist found everything equally terrifying. The principles of electricity, he knew, made him and his horse perfect conductors for the electric charge. The river was, however, as threatening. Its current was still

The Land between the Rivers

lazy enough, and the water was shallow. But the bottom was completely uncertain. The horse repeatedly stumbled, fell in deep holes, dunking both beast and rider. Then the hapless animal became stuck in quicksand, which unmercifully sucked in both man and horse, their cries for help drowned out by the sounds of the storm. Fortunately, the hunter Lee had developed a protective attitude toward his inexperienced companion. He was ever watchful for Nuttall's safety. Nuttall soon felt the rope and heard Lee's command to tie it around the horse's neck. With difficulty Lee pulled them to safety.

But could he save them from the storm? More important, could he save them from their pursuers, the two Osage warriors who had been stealthily following them all afternoon? When night and storm simultaneously fell, Lee and Nuttall scarcely knew whether or not the warriors still followed, waiting for the chance to steal or worse, kill. Lee had trapped the Arkansas valley enough to know that the Osages were capable of anything. He took nothing for granted, did everything he could to escape, and even thanked heaven for the thunderstorm, which if it terrified the pursued doubtless terrified the pursuers as well. The storm protected even as it threatened.

Recalling this experience of flight during a thunderstorm at night on the Arkansas River, Nuttall declared it to be "the most gloomy and disagreeable situation I ever experienced in my life." Upon the cool front's passing, the north wind brought a particular chill as it moved over the water. Nuttall and Lee, resting but still wet, their leather shirts and breeches refusing to dry, hid in the damp woods on the banks of the Arkansas. Nuttall was shivering, weak, still anxious. Convinced he might "perish with cold," he persuaded "Lee to kindle me a handful of fire," which was no mean achievement after a drenching storm. Lee found dry tinder and used the flint he carried to engender sparks and so to build a small blaze. But if it warmed, it could also serve as a "beacon." Refusing to give in to the temptation of such warmth, Lee stayed out of sight, while Nuttall lay next to the fire, prey to whatever enemy might be lurking in the darkness. No one approached, however: "I lay alone for two or three

130

The Cimarron, September 1819

hours, amidst the dreary howling of wolves"—probably coyotes. Nuttall, famished, cooked "a portion of a fat buck elk, which my companion had contrived to kill in the midst of our flight." The resourceful Lee joined the scientist as dawn approached, feasting on elk. They continued their journey by the fading moonlight.

Dawn brought a glorious morning, the sun glistening on the water, the trees on the bluffs enclosing the river displaying a rich verdure after the night rain, the air cool and refreshing. The rising river and wet shoreline meant that Nuttall was having a difficult time keeping up with Lee. The two men agreed to part, Lee continuing by canoe, Nuttall following the river the best he could. Lee gave Nuttall some of the elk and some flint for fire, but by day's end the elk was spoiled and the flint uncooperative, at least in Nuttall's inexperienced hands. "My gun," he noted, "was also become useless, all the powder having got wet by last night's adventure."

For two days Thomas Nuttall rode slowly south and east, paralleling and sometimes crossing the meandering Arkansas River. The weather was sunny and hot. He lived off a small amount of cooked elk but could not kindle fire nor use his gun to acquire more meat. At night he slept on the damp ground, which added to his feelings of weakness and exhaustion. His joints ached and muscles cramped under the strain of illness and privation. His legs and feet were becoming terribly swollen. On September 14, overwhelmed by the "hot and cheerless" sand, Nuttall set out east from the river, hoping to arrive at the Verdigris by land. The cane, brambles, and giant stalks of ragweed hindered travel. After several miles horse and rider came to higher ground dotted with oaks, hickories, willows, and cottonwoods. The scientist was too ill to botanize, study, collect, and observe. His chief goal was simply to survive. Ultimately Nuttall arrived, "to my great satisfaction, into the open prairies, from whence, in an elevated situation, I immediately recognized the Verdigris River."

Yet many miles still lay before him to reach Three Forks. He spent the night with the Verdigris barely in sight through the wall of

The Land between the Rivers

"an impenetrable thicket," reposing "in the rank weeds, amidst musquetoes, without fire, food, or water." The same will and determina-
tion that had driven Thomas Nuttall from the Ohio to the Mississippi to the Arkansas and Three Forks, from October 1818 to
September 1819, now compelled him to ride on one last day to
Bogy's trading post, where he congratulated himself on mere survival.

17 Fort Smith, October 1819

Thomas Nuttall returned to Three Forks a different man than when he had set out five weeks earlier. He was worn-out, ill; his body, swollen by exposure to the extremes of heat and cold, was racked by terrible spasms. He could hardly walk; indeed, for a time he felt and looked much older than his thirty-three years.

Images of laughing French boatmen, calling him "le fou," came back to haunt Nuttall. Perhaps he had been a fool. As an explorer Nuttall had successfully traveled the Ohio, Mississippi, Missouri, and Arkansas Rivers. He had crossed the Appalachians on numerous occasions. His journeys had covered thousands of miles, ranging from New England to the Great Lakes, from the upper Missouri to New Orleans, from Massachusetts to Florida. Such success had made him confident that the landscape of the unknown Arkansas Territory could likewise be conquered. His goal had been the Rockies, until the wilderness of Oklahoma got in the way. The land had conquered him, as it would so many others, and the ebullient young scientist returned from his journey beaten and half-dead.

Nuttall stayed at Bogy's trading post for a week, trying to recover his strength in body and mind. Besides fever and cramps, his mind was affected by a "horrific delirium, which perpetually dwelt upon the scene of past sufferings": ruminations of failure, obsessions about death, recollections of pain, dread of the future. Knowing he could not stay at Three Forks indefinitely, Nuttall hired a French boatman to take him downriver to Fort Smith, which would have more pleasant accommodations, as well as the watchful care of his friend Dr. Russell.

Upon arriving at Fort Smith at the end of September, Nuttall was shocked to find that "the ague and bilious [yellow] fever" had taken

The Land between the Rivers

hold of the garrison. Dr. Russell was dead. Many soldiers had been sick, too. A missionary on his way to minister to the Osages had found his mission end at Fort Smith. And Nuttall heard "that not less than 100 of the Cherokees, settled contiguous to the banks of the Arkansa, died this season of the bilious fever."

Nuttall recovered at the garrison until mid-October, when he departed once again upon the Arkansas, journeying quickly downstream to the Cadron, a corruption of "Quadrant, a name applied to the neighboring creek by the French hunters, probably in commemoration of some observation made there by that instrument, to ascertain latitude." During January and February 1820 Nuttall joined passage down the river on the boat of Henry Barbour, a merchant who had a trading post at Three Forks. At Arkansas Post Nuttall gained passage to the Mississippi with another merchant, Frederick Notrebe. Nuttall traveled by flatboat down the Mississippi to New Orleans, arriving on February 18. At New Orleans he took passage for the East Coast and Philadelphia.

Postscript

Thomas Nuttall never returned to the Arkansas River, to Three Forks, to Oklahoma. Yet more journeys to other parts of America, north and south, awaited Thomas Nuttall. While he served as a professor at Harvard College, Nuttall's restlessness took him on many smaller journeys, one of the most significant being to the White Mountains of New Hampshire, where he ascended Mount Washington to study its unique alpine climate and beautiful multiplicity of unusual flora. Eventually the botanist grew tired of teaching. He joined the expedition of Nathaniel Wyeth traversing the Rockies, from spring to fall 1834, on the Oregon Trail. The young ornithologist John Kirk Townsend accompanied the middle-aged Thomas Nuttall, the two happily losing themselves in botanical outings along the trail. At the end of the trail Nuttall made several excursions, including two voyages from 1834 to 1836 to the Sandwich Islands (Hawaii). Eventually Nuttall took passage on the *Alert* for Cape Horn, the Atlantic, and north to Boston. On board the ship was a former student, Richard Henry Dana, Jr., author of *Two Years before the Mast.*

Never again, after that dismal night on the Arkansas, did Nuttall see Mr. Lee, who drops unheralded from the pages of history. Perhaps Lee met his fate at the hands of Indian warriors or succumbed that fall of 1819 to the yellow fever. Or perhaps he lived on to see more travelers, scientists, and soldiers—the likes of Washington Irving, Stephen Long, and Sam Houston—enter the Arkansas Territory and journey up the Red, Canadian, Cimarron, and Arkansas Rivers. Perhaps Lee was on hand when Fort Gibson was established in 1824 at Three Forks.

Certainly the likes of the hunter Lee vanished over the course of

The Land between the Rivers

the 1800s from Indian Territory. When the land runs began in 1889, settlers came to one of the last American frontiers: Oklahoma. After the turn of the century oil brought people to Oklahoma and drove its economy. The land gave reluctantly to its inhabitants. Wind, drought, and erosion buried parts of Oklahoma in dust during the 1930s. After World War II the Army Corps of Engineers engaged in massive flood-control projects, building dams to generate electricity and to spur tourism. The damming of the Red River produced Lake Texoma; of the Canadian River, Lake Eufaula; of the Cimarron and Arkansas Rivers, Lake Keystone; of the Grand River, the Grand Lake o' the Cherokees, Spavinaw Lake, and Fort Gibson Lake. The Verdigris River was formed into the Kerr-McClellan Navigation System; joined with the Arkansas at Three Forks, it became the Arkansas Navigation System, a waterway to the Mississippi and the Gulf of Mexico. The Oklahoma Lake Country, as it is called, would have astonished Thomas Nuttall and Mr. Lee.

Today, drive along a desolate Oklahoma highway, experience the terror of a May thunderstorm, feel the oppressive heat of August and the biting January wind that sweeps across the plains to know what Professor Nuttall and the mysterious Mr. Lee knew. Such a land demands endurance, patience, and a certain kind of courage. Such is the heritage of the land between the rivers.

Essay on Sources

The chief sources with which to reconstruct the life, journeys, and thought of Thomas Nuttall are his journals and books, the most important being *A Journal of Travels into the Arkansa Territory, During the Year 1819: With Occasional Observations on the Manners of the Aborigines* (Philadelphia: Thom. Palmer, 1821). Nuttall dedicated the book to Zaccheus Collins, William Maclure, John Vaughan, and Joseph Corea de Serra, members of the American Philosophical Society and Academy of Natural Sciences and sponsors of his journey to the Arkansas River. I use the edition edited by Reuben Gold Thwaites, in his *Early Western Travels: 1748–1846* (Cleveland: Arthur Clark, 1905), volume 13. I also find useful the comments in the edition of Nuttall's *Journal* edited by Savoie Lottinville (Norman: University of Oklahoma Press, 1980). Nuttall's 1810 journal, reproduced with editorial comments in Jeannette Graustein's "Nuttall's Travels into the Old Northwest," *Chronica Botanica* 14 (1951): 1–88, forms the basis for chapter 5, on the Michigan Territory. The *Chronica Botanica* journal is based on a copy at the Gray Herbarium, used by permission. Another journal that provides insights into Nuttall's character is John Kirk Townsend's *Narrative of a Journey Across the Rocky Mountains to the Columbia River*, reproduced in Thwaites, *Early Western Travels*, volume 21. One of the best of Nuttall's many publications is *A Manual of the Ornithology of the United States and of Canada*, volume 1, *The Land Birds*, and volume 2, *The Water Birds* (Cambridge and Boston: Hilliard and Brown, and Hilliard, Gray, and Co., 1832, 1834).

Part of the art of historical re-creation is the use of empathy and indirect, circumstantial evidence to close the gaps not otherwise filled by the historical record. I use this subjective technique to re-create the life of Mr. Lee (chap. 7), information about whom is

The Land between the Rivers

found only in Nuttall's Arkansas journal and nowhere else. Other contemporary works inform us about the lives, tasks, customs, and characters of the American hunters and trappers who were often employed as guides for scientists and adventurers in early America. I see quite a few similarities between Mr. Lee, and his relationship with the scientist Thomas Nuttall, and the frontiersman and hunter John Evans, and his relationship with the scientist Jeremy Belknap. See my *Passaconaway's Realm: Captain John Evans and the Exploration of Mount Washington* (Hanover: University Press of New England, 2002) for the story of Captain John Evans leading Jeremy Belknap, Manasseh Cutler, and other scientists to the White Mountains and up Mount Washington. The English botanist John Bradbury's *Travels in the Interior of America, in the Years 1809, 1810, and 1811,* reprinted in Thwaites, *Early Western Travels,* volume 5, indirectly portrays Nuttall's ascent of the Missouri (chap. 6), as well as the role of hunters as guides to men of science. Good additions to the stock of knowledge on the relations of hunters and scientists are *The Journals of Lewis and Clark,* edited by Bernard De Voto (Boston: Houghton Mifflin, 1953); *Life, Letters and Papers of William Dunbar,* edited by Eron Rowland (Jackson: Mississippi Historical Society Press, 1930—quotes from which are used by permission), which forms the basis of chapter 2; and *The Expeditions of Zebulon Montgomery Pike,* 3 volumes, edited by Elliot Coues (New York: Harper, 1895), which forms the basis of chapter 3.

The journals and writings of other botanists and naturalists help to re-create the scientific, intellectual environment of the eighteenth and nineteenth centuries. The works of the English explorer, physician, and botanist John Josselyn, *New-Englands Rarities Discovered* (London, 1672) and *Two Voyages to New-England* (London, 1674), are important to understand the basis of early American materia medica in folklore and Indian knowledge of flora. The Philadelphia botanist John Bartram, in *Observations on the Inhabitants, Climate, Soil, Rivers, Productions, Animals, and Other Matters Worthy of Notice* (London, 1751), illustrates his work as one of the first great botanists/explorers in American history. There are many

Essay on Sources

essays on flora, fauna, and natural history published in the *Transactions of the American Philosophical Society* (Philadelphia: Robert Aitken, 1786), volume 2. The journal of the southern botanist William Dunbar, already cited, describes the observations of a pioneer botanist in the Louisiana Territory. The writings and journals of Manasseh Cutler, the great New England botanist, reproduced in *Life, Journals, and Correspondence of Rev. Manasseh Cutler, LL.D.*, 2 volumes, edited by William P. Cutler and Julia P. Cutler (Cincinnati: Robert Clarke and Co., 1888), include Cutler's journals of his two expeditions up Mount Washington in 1784 and 1804. Cutler provided botanical information for his friend Jeremy Belknap to include in the latter's *The History of New Hampshire* (Boston: Belknap and Young, 1792), volume 3. The journal of Nuttall's contemporary and acquaintance Jacob Bigelow, "Some Account of the White Mountains of New Hampshire," *The New England Journal of Medicine and Surgery* 5 (January 1816), is a full botanical treatment of the flora of one of the most exciting places in America for botanists to explore and to collect specimens.

Land between the Rivers also relies on other primary sources of contemporaries of Thomas Nuttall. These include Thomas Jefferson's "Notes on Virginia," in *The Life and Selected Writings of Thomas Jefferson*, edited by Adrienne Koch and William Peden (New York: Modern Library, 1944); Estwick Evans's *A Pedestrious Tour, of Four Thousand Miles, Through the Western States and Territories, during the Winter and Spring of 1818*, in Thwaites, *Early Western Travels*, volume 8; André Michaux's *Journal*, in Thwaites, *Early Western Travels*, volume 3; Fortescue Cuming's *Sketches of a Tour of the Western Country*, in Thwaites, *Early Western Travels*, volume 4; *The Journal of Jacob Fowler, Narrating an Adventure from Arkansas Through the Indian Territory, Oklahoma, Kansas, Colorado, and New Mexico to the Source of Rio Grande del Norte, 1821–22*, edited by Elliott Coues (New York: Harper, 1898); Henry Marie Brackenridge's *Journal of a Voyage up the River Missouri Performed in Eighteen Hundred and Eleven*, in Thwaites, *Early Western Travels*, volume 4; Jedidiah Morse's *Report to the Secretary of War on Indian Affairs, Comprising a Narrative of a Tour in the Summer of*

The Land between the Rivers

1820 (New Haven, 1822); and John Filson's *The Discovery, Settlement and Present State of Kentucke* (Wilmington, 1784).

For the life and work of Thomas Nuttall there is nothing better than Jeannette E. Graustein, *Thomas Nuttall, Naturalist: Explorations in America, 1808–1841* (Cambridge: Harvard University Press, 1967). I have found Graustein's book particularly helpful in re-creating some of Nuttall's lesser journeys, such as to Delaware and upstate New York in 1809 and through the Cumberland Gap in 1816. Also useful is Graustein's "Early Scientists in the White Mountains," *Appalachia,* new series, 30 (1964): 44–63. The context of Nuttall's journey among other explorers and botanists of the Arkansas valley is discussed in Stan Hoig, *Beyond the Frontier: Exploring the Indian Country* (Norman: University of Oklahoma Press, 1998); George J. Goodman and Cheryl A. Lawson, *Retracing Major Stephen H. Long's 1820 Expedition: The Itinerary and Botany* (Norman: University of Oklahoma Press, 1995); George H. Odell, *La Harpe's Post: A Tale of French-Wichita Contact on the Eastern Plains* (Tuscaloosa: University of Alabama Press, 2002); and James L. Reveal, "A Nomenclatural Morass," in *Discovering Lewis and Clark* (http://www.lewis-clark.org, 2000).

Other good treatments of early American science, exploration, and frontier life include William Goetzmann, *New Lands, New Men: America and the Second Great Age of Discovery* (New York: Penguin Books, 1986); John C. Greene, *American Science in the Age of Jefferson* (Ames: Iowa State University Press, 1984); Bernard De Voto, *The Course of Empire* (Boston: Houghton Mifflin, 1980); David Weber, *The Spanish Frontier in North America* (New Haven: Yale University Press, 1992); Malcolm J. Rohrbough, *The Trans-Appalachian Frontier: People, Societies, and Institutions, 1775–1850* (New York: Oxford University Press, 1978); John S. Otto, *The Southern Frontiers: The Agricultural Evolution of the Colonial and Antebellum South* (Westport: Greenwood Press, 1989); John Mack Faragher, *Daniel Boone: The Life and Legend of an American Pioneer* (New York: Henry Holt, 1992); Richard C. Wade, *The Urban Frontier: Pioneer Life in Early Pittsburgh, Cincinnati, Lexington, Louisville, and St. Louis* (Chicago: University of Chicago Press, 1959); Daniel J. Boorstin, *The Lost World of Thomas Jefferson* (Chicago: University of

Essay on Sources

Chicago Press, 1981), as well as Boorstin's *The Discoverers* (New York: Random House, 1983); Alvin M. Josephy Jr., *The Indian Heritage of America* (Boston: Houghton Mifflin, 1991); Russell M. Lawson, "Science and Medicine," in *American Eras: The Colonial Era* (Detroit: Gale Publishing, 1998); and *Louisiana Purchase: An Encyclopedia*, edited by Junius Rodriguez (Santa Barbara: ABC-CLIO, 2002).

Useful regional studies include S. Charles Bolton, *Arkansas, 1800–1860, Remote and Restless* (Fayetteville: University of Arkansas Press, 1998); Carl N. Tyson, *The Red River in Southwestern History* (Norman: University of Oklahoma Press, 1981); H. Wayne Morgan and Anne Hodges Morgan, *Oklahoma: A History* (New York: W. W. Norton, 1984); Tom Meagher, *Sketch Map of the Three Forks* (Tulsa, 1942); and the many articles found in the *Chronicles of Oklahoma*, the journal of the Oklahoma Historical Society.

My knowledge of the flora and fauna of the land between the rivers has been informed by the many useful Audubon and Peterson field guides, for example, John Bull and John Farrand Jr., *National Audubon Society Field Guide to North American Birds, Eastern Region* (New York: Knopf, 1994); Elbert L. Little, *National Audubon Society Field Guide to Trees, Eastern Region* (New York: Knopf, 1980); William A. Niering, John W. Thieret, and Nancy C. Olmstead, *National Audubon Society Field Guide to North American Wildflowers, Eastern Region* (New York: Knopf, 2001); Ann Sutton and Myron Sutton, *Eastern Forests* (New York: Knopf, 1985); Stephen Whitney, *Western Forests* (New York: Knopf, 1985); Steven Foster and James A. Duke, *A Field Guide to Medicinal Plants and Herbs of Eastern and Central North America* (New York: Houghton Mifflin, 2000); and Steven Foster and Roger A. Caras, *A Field Guide to Venomous Animals and Poisonous Plants: North America North of Mexico* (New York: Houghton Mifflin, 1994). I have supplemented such works with Janine M. Benyus, *The Field Guide to Wildlife Habitats of the Western United States* (New York: Simon and Schuster, 1989), and *Botanica's Trees and Shrubs* (San Diego: Laurel Glen Publishing, 1999). Whenever choosing between Nuttall's scientific nomenclature and the modern nomenclature of field guides, I have used the latter.

Index

143

Index

Index

Index

Index

147

Index

Index

Index

Index

Index

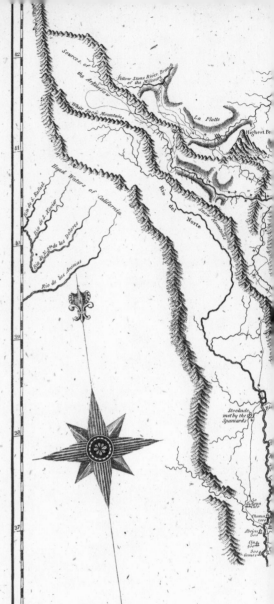

42

41

40

39

38

37

36

35

34

Sources or
the Arkansaw

Yellow Stone River, Branch
of the Missouri

White Snow Mountains

La Platte

Highest Pe

High Waters   of   California

Rio de S. Rafael

Rio de S. Xavier

Rio del   Norte

Rio de N. S.ᵃ de las Dolores

Snow Mountains

Rio de las Animas

Stockade
met by the
Spaniards

So Juan
Pueblo

Chama
tove

Abiguú
Soo
Oso
Gomez Soo
Soo

REFERENCE.

———————— Route of the American Exploring Party.

════════════ Route pursued by the Spaniards going out.

──────────── Excursions made by Capt.ⁿ Pike with 2 or 3 men.

× ×   American Camps.

○ ○   Spanish Camps.

▦   Limits of actual Surveys.

⚲   Spanish Villages or Towns.

△   Indian Villages.

The Red River to A was surveyed by Tho.ˢ Freeman Esq.ʳ

The Whashila to B.   „   „   „   W.ᵐ Dunbar Esq.ʳ

The Arkansaw from C to its mouth, by Lieu.ᵗ Wilkinson U.S.Inf.ʸ

The White River to D by Captain Many U.S. Art.ʸ

The Missouri from the mouth of the Osage to the
    entrance of La Platte by Captain M. Lewis.

The Mississippi from Bradley's Map.